# 2015: An Astronomical Year

*A Reference Guide to 365 Nights of Astronomy*

Richard J. Bartlett

Copyright © 2014 by Richard J. Bartlett

Published by Stars 'n Stuff Publishing

All rights reserved. Except for brief excerpts used in a review, this book or any portion thereof may not be reproduced or used in any manner whatsoever without the express written permission of the publisher.

First Edition, September 2014

# Contents

2015: An Astronomical Year ..................................................... 1
Introduction ............................................................................ 6
    About the Author ............................................................... 6
    Why I Wrote This Book ..................................................... 6
    Who This Book Is For ........................................................ 7
    How to Use This Book ....................................................... 7
    Acknowledgements ............................................................ 9
    Also by the Author ............................................................ 9
    The Author Online ........................................................... 10
2015 ..................................................................................... 11
    Solar System Summary ................................................... 13
    January ............................................................................ 22
    February .......................................................................... 29
    March .............................................................................. 34
    April ................................................................................ 41
    May .................................................................................. 48
    June ................................................................................. 55
    July .................................................................................. 62
    August ............................................................................. 70
    September ....................................................................... 77
    October ........................................................................... 84
    November ........................................................................ 91
    December ........................................................................ 98

# Introduction

About the Author
I've had an interest in astronomy since I was six and although my interest has waxed and waned like the Moon, I've always felt compelled to stop and stare at the stars.

In the late 90's, I discovered the booming frontier of the internet, and like a settler in the Midwest, I quickly staked my claim on it. I started to build a (now-defunct) website called *StarLore*. It was designed to be an online resource for amateur astronomers who wanted to know more about the constellations - and all the stars and deep sky objects to be found within them. It was quite an undertaking.

After the website was featured in the February 2001 edition of *Sky & Telescope* magazine, I began reviewing astronomical websites and software for their rival, *Astronomy*. This was something of a dream come true; I'd been reading the magazine since I was a kid and now my name was regularly appearing in it.

Unfortunately, a financial downturn forced my monthly column to be cut after a few years but I'll always be grateful for the chance to write for the world's best-selling astronomy magazine.

I emigrated from England to the United States in 2004 and spent three years under relatively clear, dark skies in Oklahoma. I then relocated to Kentucky in 2008 and then California in 2013. I now live in the suburbs of Los Angeles; not the most ideal location for astronomy, but there are still a number of naked eye events that are easily visible on any given night.

Why I Wrote This Book
I published my first astronomical Kindle eBook, *2015 An Astronomical Year* on Amazon, July 16th 2014. I then made it available as a free

download for five days and got a number of requests for a paperback version from non-Kindle readers.

This edition is a result of those requests.

The original eBook contained graphics, with accompanying explanations, depicting highlighted astronomical events for absolute beginners. Unfortunately, due to printing costs and some concerns about how those images might appear in print, I decided to omit those graphics and publish a purely textual version instead.

Who This Book Is For
This book is not for the absolute beginner. Again, if you're new to the hobby, I would recommend the electronic version of this book (see *Why I Wrote This Book* above and *Also by the Author* below) or any one of the other astronomy books available.

If, however, you have a little experience and know your conjunctions from your oppositions and your magnitude from your apparent diameter, then this book is probably just fine for you.

How to Use This Book
The book is divided into five sections, one for each year. Each section begins with a summary of moon phases, eclipses, planetary visibility, notable conjunctions and meteor showers for the year.

The book then details each individual month with notes on what's happening and what's visible on each of the key dates in that month. The text is designed to be short and to the point and aims to provide the most relevant information in the least amount of time.

Here's an example:

## January 11th

*01:00 UT - Conjunction of Mercury and Venus. Separation: 38.5'. (Mercury: 71% illuminated, magnitude -0.7, diameter 6.2". Venus: 95% illuminated, magnitude -3.9, diameter 10.5". Capricornus, evening sky.)*

All the times are given in Universal Time (UT), the standard "time zone" used in the astronomical community. If you live in the United Kingdom, then the only conversion you'll need to add is an hour for British Summer Time.

Unfortunately, for the rest of the world, it's just not practical to convert all the times into the different time zones that are in force across the globe, but you can find an easy conversion tool at the following URL: http://www.worldtimeserver.com/convert_time_in_UTC.aspx

Next is a description of the event itself with details given for each planet or asteroid involved. You'll note that I don't specify the angular separation for conjunctions involving the Moon; this is because the Moon's separation from a star or planet will differ depending upon your latitude. Therefore, for example, the Moon will often appear to occult Aldebaran from some locations but not from others.

Bear in mind that the timing of lunar occultations are also subject to your longitude as the event may take place during daylight at your location. So even though the Moon might be visible in the evening sky, it may have moved on slightly from the time of the event itself. (It will still be worth observing but the conjunction may not appear as close.)

Fortunately, planets move slowly enough to make purely planetary conjunctions observable for several days, both before and after the date of closest approach.

At the end of the description the text will tell you if the object(s) are visible in the pre-dawn or evening sky or not visible at all. This is based upon the visibility from the equator and the visibility from your own location will depend upon your latitude. However, the text will hopefully give you some indication of when it may be possible to make an observation.

Acknowledgements

The vast majority of the data in this book comes from a single source, the *Mobile Observatory* app for Android devices by Wolfgang Zima. I bought this excellent app a few years ago and I've used it almost daily ever since. Quite honestly, I'd recommend it to anyone and everyone with an Android device and I want to thank Wolfgang for allowing me to use the data. You can download it by searching for Mobile Observatory from the Google Play store or by visiting zima.co online.

I'd also like to thank the members of the *Telescope Addicts – Astronomy & Astrophotography Community* on Facebook. It was their generous support that helped to make *2015 An Astronomical* year such a success in the first place.


Also by the Author

*2015 An Astronomical Year* is an astronomical eBook for everyone that details hundreds of night sky events throughout the year. It was designed for both the general public (with highlighted text and graphics to showcase the best events) and astronomers of all levels. It includes details of the lunar phases and eclipses, as well as conjunctions, oppositions, magnitude and apparent diameter changes for the planets and major asteroids. (With thousands of downloads within the first few months alone, it has ranked #1 in Free Kindle Astronomy books, within the Top 10 Paid Kindle Astronomy books and within the Top 50 Free Kindle Non-Fiction books.)

*The Astronomical Almanac (2015-2019): A Comprehensive Guide to Night Sky Events* provides details of thousands of astronomical events from 2015 through to 2019. Designed for astronomers of all levels, this the guide includes almost daily data and information on the Moon and planets, as well as Pluto, Ceres, Pallas, Juno and Vesta. It was downloaded over 5,000 times during its first six weeks, ranked #1 in the Free Kindle Astronomy book category, #3 in the Paid Kindle Astronomy book category and within the Top 50 of *all* Free Kindle books in October 2014. (Available as an eBook worldwide and as a paperback from the US or UK.)

*The Amateur Astronomer's Notebook: A Journal for Recording and Sketching Astronomical Observations* is the perfect way to log your observations of the Moon, stars, planets and deep sky objects. With an additional appendix with hundreds of suggested deep sky objects, this 8.5" by 11" notebook allows you to record everything you need for 150 observing sessions under the stars. (Paperback only and available from the US or UK.)

*Echoes of Earth* – a collection of science fiction, mythological and philosophical short stories that I wrote many, many moons ago. (i.e, in the mid 1990's.)


The Author Online
Email: astronomywriter@gmail.com

Amazon:
http://www.amazon.com/Richard-J.-Bartlett/e/B00LY84PV2/

The Astronomical Year: http://theastronomicalyear.wordpress.com/

Facebook:
https://www.facebook.com/pages/Richard-J-Bartlett/1506578736250881

Twitter (@astronomywriter): https://twitter.com/astronomywriter


Clear skies!

Richard J. Bartlett

September, 2014

# 2015

# *Solar System Summary*

## *Moon*

### Lunar Phases

|  | Jan | Feb | Mar | Apr | May | Jun | Jul | Aug | Sep | Oct | Nov | Dec |
|---|---|---|---|---|---|---|---|---|---|---|---|---|
| 1st |  |  |  |  |  |  |  |  |  |  |  |  |
| 2nd |  |  |  |  |  | ○ | ○ |  |  |  |  |  |
| 3rd |  | ○ |  |  |  |  |  |  |  |  | ◐ | ◐ |
| 4th |  |  |  | ○ | ○ |  |  |  |  | ◐ |  |  |
| 5th | ○ |  | ○ |  |  |  |  |  | ◐ |  |  |  |
| 6th |  |  |  |  |  |  |  |  |  |  |  |  |
| 7th |  |  |  |  |  |  |  | ◐ |  |  |  |  |
| 8th |  |  |  |  |  |  |  | ◐ |  |  |  |  |
| 9th |  |  |  |  |  | ◐ |  |  |  |  |  |  |
| 10th |  |  |  |  |  |  |  |  |  |  |  |  |
| 11th |  |  |  | ◐ |  |  |  |  |  |  | ● | ● |
| 12th |  | ◐ |  | ◐ |  |  |  |  |  |  |  |  |
| 13th | ◐ |  | ◐ |  |  |  |  |  | ● | ● |  |  |
| 14th |  |  |  |  |  |  |  | ● |  |  |  |  |
| 15th |  |  |  |  |  |  |  |  |  |  |  |  |

# Lunar Phases (cont)

|      | Jan | Feb | Mar | Apr | May | Jun | Jul | Aug | Sep | Oct | Nov | Dec |
|------|-----|-----|-----|-----|-----|-----|-----|-----|-----|-----|-----|-----|
| 16th |     |     |     |     |     | ●   | ●   |     |     |     |     |     |
| 17th |     |     |     |     |     |     |     |     |     |     |     |     |
| 18th |     | ●   |     | ●   | ●   |     |     |     |     |     |     | ◐   |
| 19th |     |     |     |     |     |     |     |     |     |     | ◐   |     |
| 20th | ●   |     | ●   |     |     |     |     |     | ◐   |     |     |     |
| 21st |     |     |     |     |     |     |     | ◐   |     |     |     |     |
| 22nd |     |     |     |     |     |     |     | ◐   |     |     |     |     |
| 23rd |     |     |     |     |     |     |     |     |     |     |     |     |
| 24th |     |     |     |     |     | ◐   | ◐   |     |     |     |     |     |
| 25th |     | ◐   |     | ◐   | ◐   |     |     |     |     |     | ○   | ○   |
| 26th |     |     |     |     |     |     |     |     |     |     |     |     |
| 27th | ◐   |     | ◐   |     |     |     |     |     |     | ○   |     |     |
| 28th |     |     |     |     |     |     |     |     | ○   |     |     |     |
| 29th |     |     |     |     |     |     |     | ○   |     |     |     |     |
| 30th |     |     |     |     |     |     |     |     |     |     |     |     |
| 31st |     |     |     |     |     | ○   |     |     |     |     |     |     |

# *Eclipses*

## Lunar Eclipses

April 4th - 12:01 UT – Total lunar eclipse. Visible from the Antarctica, the Arctic, most of Asia, the Atlantic, Australia, the Indian Ocean, most of North and South America and the Pacific.

September 28th - 02:47 UT – Total lunar eclipse. Visible from Africa, Antarctica, the Arctic, south and east Asia, the Atlantic, Europe, the Indian Ocean, most of North and South America and the Pacific.

## Solar Eclipses

March 20th - 09:45 UT – Total solar eclipse. Visible from north and west Africa, the Arctic, north and east Asia, the Atlantic, Europe and western North America.

September 13th - 06:55 UT – Partial solar eclipse. Visible from southern Africa, Antarctica, the Atlantic and the Indian Ocean.

# Planet Visibility

## January to April

|  |  | Mer | Ven | Mar | Jup | Sat | Ura | Nep | Plu | Cer | Pal | Jun | Ves |
|---|---|---|---|---|---|---|---|---|---|---|---|---|---|
| **January** | Early |  |  |  |  |  |  |  | C |  |  |  |  |
|  | Mid | E |  |  |  |  |  |  |  |  |  |  |  |
|  | Late |  |  |  |  |  |  |  |  |  |  | O |  |
| **February** | Early |  |  |  | O |  |  |  |  |  |  |  |  |
|  | Mid |  |  |  |  |  |  |  |  |  |  |  |  |
|  | Late | E |  |  |  |  |  | C |  |  |  |  |  |
| **March** | Early |  |  |  |  |  |  |  |  |  |  |  |  |
|  | Mid |  |  |  |  |  |  |  |  |  |  |  |  |
|  | Late |  |  |  |  |  |  |  |  |  |  |  |  |
| **April** | Early |  |  |  |  |  | C |  |  |  |  |  |  |
|  | Mid |  |  |  |  |  |  |  |  |  |  |  |  |
|  | Late |  |  |  |  |  |  |  |  |  |  |  |  |

### Key

|  |  |  |  |
|---|---|---|---|
| | Evening Sky | C | In conjunction with the Sun |
| | Pre-dawn sky | E | Greatest elongation (Mercury & Venus only) |
| | Not Visible | O | Opposition with the Sun. (Outer planets and asteroids only) |

# May to August

| Month | Period | Mer | Ven | Mar | Jup | Sat | Ura | Nep | Plu | Cer | Pal | Jun | Ves |
|---|---|---|---|---|---|---|---|---|---|---|---|---|---|
| May | Early | E | | | | | | | | | | | |
| May | Mid | | | | | | | | | | | | |
| May | Late | C | | | | O | | | | | | | C |
| June | Early | | E | | | | | | | | | | |
| June | Mid | | | C | | | | | | | O | | |
| June | Late | E | | | | | | | | | | | |
| July | Early | | | | | | | | O | | | | |
| July | Mid | | | | | | | | | | | | |
| July | Late | C | | | | | | | | O | | | |
| August | Early | | | | | | | | | | | | |
| August | Mid | | C | | | | | | | | | | |
| August | Late | | | | C | | | | | | | | |

| Key | | | | | |
|---|---|---|---|---|---|
| | | Evening Sky | | C | In conjunction with the Sun |
| | | Pre-dawn sky | | E | Greatest elongation (Mercury & Venus only) |
| | | Not Visible | | O | Opposition with the Sun. (Outer planets and asteroids only) |

17

## September to December

|  |  | Mer | Ven | Mar | Jup | Sat | Ura | Nep | Plu | Cer | Pal | Jun | Ves |
|---|---|---|---|---|---|---|---|---|---|---|---|---|---|
| September | Early | E |  |  |  |  |  | O |  |  |  |  |  |
| September | Mid |  |  |  |  |  |  |  |  |  |  |  |  |
| September | Late | C |  |  |  |  |  |  |  |  |  | C |  |
| October | Early |  |  |  |  |  |  |  |  |  |  |  |  |
| October | Mid | E |  |  |  |  | O |  |  |  |  |  |  |
| October | Late |  | E |  |  |  |  |  |  |  |  |  |  |
| November | Early |  |  |  |  |  |  |  |  |  |  |  |  |
| November | Mid | C |  |  |  |  |  |  |  |  |  |  |  |
| November | Late |  |  |  |  | C |  |  |  |  |  |  |  |
| December | Early |  |  |  |  |  |  |  |  |  |  |  |  |
| December | Mid |  |  |  |  |  |  |  |  |  |  |  |  |
| December | Late | E |  |  |  |  |  |  |  |  |  |  |  |

| Key | | Evening Sky | | | C | In conjunction with the Sun |
|---|---|---|---|---|---|---|
|  |  | Pre-dawn sky |  |  | E | Greatest elongation (Mercury & Venus only) |
|  |  | Not Visible |  |  | O | Opposition with the Sun. (Outer planets and asteroids only) |

# *Notable Conjunctions*

## Pre-dawn sky

*March 4th* - 13:00 UT – Mercury is 54' north of asteroid 4 Vesta. (Mercury: 70% illuminated, magnitude 0.0, diameter 6.2". Vesta: magnitude 7.6. Capricornus, pre-dawn sky.)

*May 15th* - 13:25 UT – The waning crescent Moon is north of Uranus. An occultation may be visible from some parts of the world. (Uranus: magnitude 5.9, diameter 3.4". Pisces, pre-dawn sky.)

*October 17th* - 13:44 UT – Mars is 23' north of Jupiter. (Mars: 96% illuminated, magnitude 1.7, diameter 4.1". Jupiter: magnitude -1.8, diameter 32.1", Leo, pre-dawn sky.)

*October 18th* - 19:26 UT - Mars occults the magnitude 4.6 star Chi Leo. (Mars: 96% illuminated, magnitude 1.7, diameter 4.1". Leo, pre-dawn sky.)

*November 3rd* - 07:38 UT – Venus is 41' south of Mars. (Venus: 55% illuminated, magnitude -4.3, diameter 22.1". Mars: 95% illuminated, magnitude 1.7, diameter 4.3", Virgo, pre-dawn sky.)

*December 6th* - 00:46 UT – The waning crescent Moon is north of Mars. An occultation may be visible from some parts of the world. (Mars: 93% illuminated, magnitude 1.5, diameter 4.9", Virgo, pre-dawn sky.)

## Evening sky

*January 11th* - 01:00 UT - Conjunction of Mercury and Venus. Separation: 38.5'. (Mercury: illuminated 71%, magnitude -0.7, diameter 6.2". Venus: 95% illuminated, magnitude -3.9, diameter 10.5". Capricornus, evening sky.)

*January 19th* - 21:21 UT - Mars is 13' south of Neptune. (Mars: 95% illuminated, magnitude 1.2, diameter 4.6". Neptune: magnitude 8.0, diameter 2.2". Aquarius, evening sky.)

*February 21st* - 19:53 UT – Venus is 25' south of Mars. (Venus: 88% illuminated, magnitude -4.0, diameter 11.7". Mars: 97% illuminated, magnitude 1.3, diameter 4.3". Pisces, evening sky.)

*March 4th* - 18:41 UT – Venus is 5.3' north of Uranus. (Venus: 86% illuminated, magnitude -4.0, diameter 12.2". Uranus: magnitude 5.9, diameter 3.4". Pisces, evening sky.)

*March 21st* - 10:09 UT – The waxing crescent Moon is south of Uranus. An occultation may be visible from some parts of the world. (Uranus: magnitude 5.9, diameter 3.3". Pisces, not visible.)

*July 1st* - 18:41 UT – Venus is 5.3' north of Uranus. (Venus: 86% illuminated, magnitude -4.0, diameter 12.2". Uranus: magnitude 5.9, diameter 3.4". Pisces, evening sky.)

*August 7th* - 07:34 UT - Mercury is 32' north of Jupiter. (Mercury: 88% illuminated, magnitude -0.6, diameter 5.2". Jupiter: magnitude -1.7, diameter 31.0". Leo, evening sky.)

*August 7th* - 20:23 UT - Mercury is 53' north of Regulus. (88% illuminated, magnitude -0.6, diameter 5.2". Leo, evening sky.)

## *Meteor Showers*

*January 4th* - The Quadrantid meteor shower peaks. Maximum zenith hourly rate: 120. (Moon: almost full. Boötes.)

*April 22nd* - The Lyrid meteor shower peaks. Maximum zenith hourly rate: 18. (Moon: waxing crescent. Lyra.)

*May 5th* - The Eta Aquariid meteor shower peaks. Maximum zenith hourly rate: 65. (Moon: just past full. Aquarius.)

*June 7th* - The Arietid meteor shower peaks. Maximum zenith hourly rate: 54. (Moon: waning gibbous. Aries.)

*July 29th* - The Southern Delta Aquariid meteor shower peaks. Maximum zenith hourly rate: 16. (Moon: waxing gibbous. Aquarius.)

*July 29th* - The Beta Cassiopeid meteor shower peaks. Maximum zenith hourly rate: 10. (Moon: waxing gibbous. Cassiopeia.)

*August 12th* - The Perseid meteor shower peaks. Maximum zenith hourly rate: 100. (Moon: waning crescent. Perseus.)

*October 8th* - The Draconid meteor shower peaks. Maximum zenith hourly rate: Variable. (Moon: waning crescent. Draco.)

*October 21st* - The Orionid meteor shower peaks. Maximum zenith hourly rate: 25. (Moon: just-past full. Orion.)

*November 17th* - The Leonid meteor shower peaks. Maximum zenith hourly rate: 15. (Moon: waxing crescent. Leo.)

*December 13th* - The Geminid meteor shower peaks. Maximum zenith hourly rate: 120. (Moon: waxing crescent. Gemini.)

*December 23rd* - The Ursid meteor shower peaks. Maximum zenith hourly rate: 10. (Moon: waxing gibbous. Ursa Minor.)

# *January*

### *January 1st*

12:29 UT – The waxing gibbous Moon is south of M45, the Pleiades open star cluster. (Taurus, evening sky.)

Jupiter brightens to magnitude -2.5. (Apparent diameter 43.3". Leo, pre-dawn sky.)

### *January 2nd*

11:04 UT – The waxing gibbous Moon is north of Aldebaran. (Taurus, evening sky.)

### *January 3rd*

12:42 UT - Pluto is in conjunction with the Sun. Distance to Earth: 33.774 AU (Pluto: magnitude 14.2. Sagittarius, not visible.)

Venus leaves Sagittarius and enters Capricornus. (96% illuminated, magnitude -3.9, diameter 10.4", evening sky.)

### *January 4th*

08:24 UT – The Earth is at perihelion. Distance to Sun: 0.983 AU

Asteroid 2 Pallas leaves Serpens and enters Ophiuchus. (Magnitude 9.5, pre-dawn sky.)

The Quadrantid meteor shower peaks. Maximum zenith hourly rate: 120. (Moon: almost full. Boötes.)

*January 5th*

04:54 UT - Full Moon. (Gemini, visible all night.)

19:12 UT - Pluto is at apogee. Distance to Earth: 33.775 AU. (Magnitude 14.2. Sagittarius, not visible.)

Asteroid 3 Juno brightens to magnitude 8.0. (Hydra, evening sky.)

*January 8th*

05:33 UT – The waning gibbous Moon is south of Jupiter. (Jupiter: magnitude -2.5, diameter 44.0". Leo, pre-dawn sky)

Mars leaves Capricornus and enters Aquarius. (95% illuminated, magnitude 1.1, diameter 4.7", evening sky.)

*January 9th*

00:22 UT – The waning gibbous Moon is south of Regulus. (Leo, pre-dawn sky.)

*January 10th*

19:19 UT – Asteroid 4 Vesta is in conjunction with the Sun. Distance to Earth: 3.168 AU (Magnitude 7.4. Sagittarius, not visible.)

*January 11th*

01:00 UT - Conjunction of Mercury and Venus. Separation: 38.5'. (Mercury: 71% illuminated, magnitude -0.7, diameter 6.2". Venus: 95% illuminated, magnitude -3.9, diameter 10.5". Capricornus, evening sky.)

07:39 UT - Summer begins in the southern hemisphere of Mars. (95% illuminated, magnitude 1.1, diameter 4.6". Aquarius, evening sky.)

*January 13th*

09:46 UT – Last Quarter Moon. (Virgo, pre-dawn sky.)

11:18 UT – The last quarter Moon is north of Spica. (Virgo, pre-dawn sky.)

*January 14th*

20:30 UT - Mercury reaches Greatest Eastern Elongation (57% illuminated, magnitude -0.6, diameter 6.8". Capricornus, evening sky.)

Mercury fades to magnitude -0.5. (61% illuminated, diameter 6.7". Capricornus, evening sky.)

*January 15th*

Saturn leaves Libra and enters Scorpius. (Magnitude 0.6, diameter 15.7", pre-dawn sky.)

*January 16th*

13:12 UT – The waning crescent Moon is north of Saturn (Saturn: magnitude 0.6, diameter 15.7". Scorpius, pre-dawn sky.)

14:06 UT - Mercury reaches half phase (50% illuminated, magnitude -0.4, diameter 7.2". Capricornus, evening sky.)

22:20 UT – The waning crescent Moon is north of Antares. (Ophiuchus, pre-dawn sky.)

Neptune fades to magnitude 8.0. (Diameter 2.2". Aquarius, evening sky.)

Asteroid 2 Pallas leaves Ophiuchus and enters Hercules. (Magnitude 9.4, pre-dawn sky.)

*January 17th*

18:57 UT – Asteroid 4 Vesta is at apogee. Distance to Earth: 3.170 AU (Magnitude 7.5. Sagittarius, not visible.)

Good opportunity to see Earthshine on the waning crescent Moon. (Pre-dawn sky.)

*January 18th*

19:51 UT – The waning crescent Moon is north of dwarf planet Ceres. (Ceres: magnitude 8.7. Sagittarius, pre-dawn sky.)

*January 19th*

09:00 UT - Mercury fades to magnitude 0.0 (37% illuminated, diameter 7.8". Capricornus, evening sky.)

11:00 UT – The almost-new Moon is north of Pluto. (Pluto: magnitude 14.2. Sagittarius, not visible.)

21:21 UT - Mars is 13' south of Neptune. (Mars: 95% illuminated, magnitude 1.2, diameter 4.6". Neptune: magnitude 8.0, diameter 2.2". Aquarius, evening sky.)

*January 20th*

13:15 UT - New Moon. (Sagittarius, not visible.)

18:57 UT – The just-past new Moon is north of asteroid 4 Vesta. (Vesta: magnitude 7.5. Sagittarius, not visible.)

The Sun leaves Capricornus and enters Aquarius.

Uranus fades to magnitude 5.9. (Apparent diameter 3.5". Pisces, evening sky.)

*January 21st*

03:48 UT - Mercury is stationary prior to beginning retrograde motion. (28% illuminated, magnitude 0.4, diameter 8.3". Capricornus, evening sky.)

09:42 UT – Asteroid 3 Juno is at perigee. Distance to Earth: 1.324 AU (Magnitude 7.9. Hydra, visible all night.)

17:06 UT – The just-past new Moon is north of Mercury. (Mercury: 26% illuminated, magnitude 0.6, diameter 8.4". Capricornus, evening sky.)

20:30 UT – Mercury is at perihelion. Distance to Sun: 0.308 AU. (25% illuminated, magnitude 0.6, diameter 8.4". Aquarius, evening sky.)

Mercury leaves Capricornus and enters Aquarius. (28% illuminated, magnitude 0.4, diameter 8.3". Capricornus, evening sky.)

Mercury fades to magnitude 0.5. (28% illuminated, diameter 8.3". Capricornus, evening sky.)

*January 22nd*

04:00 UT – The waxing crescent Moon is north of Venus. (Venus: 93% illuminated, magnitude -3.9, diameter 10.8". Capricornus, evening sky.)

Mercury fades to magnitude 1.0 (17% illuminated, diameter 8.8". Aquarius, evening sky.)

Jupiter increases its apparent diameter to 45.0". (Magnitude -2.5, Leo, evening sky.)

## *January 23rd*

01:19 UT – The waxing crescent Moon is north of Neptune. (Neptune: magnitude 8.0, diameter 2.2". Aquarius, evening sky.)

04:03 UT – The waxing crescent Moon is north of Mars. (Mars: 96% illuminated, magnitude 1.2, diameter 4.5". Aquarius, evening sky.)

Good opportunity to see Earthshine on the waxing crescent Moon. (Evening sky.)

## *January 24th*

Asteroid 2 Pallas leaves Hercules and returns to Ophiuchus. (Magnitude 9.4, pre-dawn sky.)

## *January 25th*

09:32 UT – Asteroid 3 Juno reaches its maximum brightness. Magnitude: 7.9 (Hydra, visible all night.)

10:07 UT – The waxing crescent Moon is north of Uranus. (Uranus: magnitude 5.9, diameter 3.4". Pisces, evening sky.)

Venus leaves Capricornus and enters Aquarius. (93% illuminated, magnitude -3.9, diameter 10.9", evening sky.)

Saturn increases its apparent diameter to 16.0". (Magnitude 0.6, Scorpius, pre-dawn sky.)

*January 26th*

16:02 UT – Asteroid 3 Juno is at opposition. (Magnitude 7.9. Hydra, visible all night.)

*January 27th*

04:48 UT - First Quarter Moon. (Aries, evening sky.)

*January 28th*

Asteroid 4 Vesta leaves Sagittarius and enters Capricornus. (Magnitude 7.5, not visible.)

*January 29th*

18:24 UT – The waxing gibbous Moon is north of Aldebaran. (Taurus, evening sky.)

*January 30th*

13:48 UT - Mercury is at inferior conjunction with the Sun. Distance to Earth: 0.659 AU. (1% illuminated, magnitude 4.6, diameter 10.1". Capricornus, not visible.)

*January 31st*

Saturn brightens to magnitude 0.5. (Apparent diameter 16.1", Scorpius, pre-dawn sky.)

# *February*

## *February 1st*

05:30 UT - Mercury is at perigee. Distance to Earth: 0.655 AU (2% illuminated, magnitude 4.1, diameter 10.2". Capricornus, not visible.)

11:22 UT – Venus is 48' south of Neptune. (Venus: 92% illuminated, magnitude -3.9, 11.1", Aquarius, evening sky. Neptune: magnitude 8.0, diameter 2.2". Aquarius, not visible.)

## *February 3rd*

07:07 UT – The almost full Moon is north of asteroid 3 Juno. (Juno: magnitude 8.0. Hydra, visible all night.)

23:09 UT - Full Moon (Cancer, visible all night)

## *February 4th*

05:41 UT – The just-past full Moon is south of Jupiter. (Jupiter: magnitude -2.6, diameter 45.3". Leo, visible all night.)

15:18 UT – Mercury is 4.9° north of asteroid 4 Vesta. (Mercury: 8% illuminated, magnitude 2.5, diameter 10.0". Vesta: magnitude 7.6. Capricornus, not visible)

## *February 5th*

05:34 UT – The waning gibbous Moon is south of Regulus. (Leo, visible all night.)

Jupiter returns to Cancer from Leo. (Magnitude -2.6, diameter 45.3". Visible all night.)

*February 6th*

07:10 UT - Jupiter at perigee: Distance to Earth: 4.346 AU (Magnitude -2.6, diameter 45.3". Cancer, visible all night.)

15:10 UT - Jupiter reaches maximum brightness. Magnitude: -2.6 (Diameter 45.3". Cancer, visible all night.)

18:20 UT - Jupiter is at opposition. (Magnitude -2.6, diameter 45.3". Cancer, visible all night.)

*February 9th*

16:18 UT – The waning gibbous Moon is north of Spica. (Virgo, pre-dawn sky.)

Mercury brightens to magnitude 1.0. (23% illuminated, diameter 9.2". Capricornus, pre-dawn sky.)

*February 11th*

06:27 UT - Mercury is stationary prior to resuming prograde motion. (28% illuminated, magnitude 0.8, diameter 8.9". Capricornus, pre-dawn sky.)

Mars leaves Aquarius and enters Pisces. (97% illuminated, magnitude 1.2, diameter 4.3", evening sky.)

*February 12th*

03:49 UT - Last Quarter Moon. (Libra, pre-dawn sky.)

22:25 UT – The last quarter Moon is north of Saturn. (Saturn: magnitude 0.5, diameter 16.4". Scorpius, pre-dawn sky.)

*February 13th*

09:14 UT – The just-past last quarter Moon is Antares. (Ophiuchus, pre-dawn sky.)

Mercury brightens to magnitude 0.5. (33% illuminated, diameter 8.6". Capricornus, pre-dawn sky.)


*February 15th*

22:45 UT – The waning crescent Moon is north of Pluto. (Pluto: magnitude 14.2. Sagittarius, pre-dawn sky.)

23:32 UT – The waning crescent Moon is north of dwarf planet Ceres. (Ceres: magnitude 8.7. Sagittarius, pre-dawn sky.)

Good opportunity to see Earthshine on the waning crescent Moon. (Pre-dawn sky.)


*February 16th*

The Sun leaves Capricornus and enters Aquarius.

Venus leaves Aquarius and enters Pisces. (89% illuminated, magnitude -3.9, diameter 11.5", evening sky.)

Asteroid 3 Juno leaves Hydra and enters Cancer. (Magnitude 8.2, evening sky.)


*February 17th*

04:34 UT – The waning crescent Moon is north of Mercury. (Mercury: 44% illuminated, magnitude 0.3, diameter 7.9". Capricornus, pre-dawn sky.)

15:18 UT – The waning crescent Moon is north of asteroid 4 Vesta. (Vesta: magnitude 7.6. Capricornus, not visible.)

Venus brightens to magnitude -4.0. (89% illuminated, diameter 11.6". Pisces, evening sky.)

*February 18th*

23:47 UT - New Moon. At a distance of just under 357,000 km (222,000 miles,) this is the nearest new moon of the year. (Aquarius, not visible.)

Jupiter decreases its apparent diameter to 45.0". (Magnitude -2.6, Cancer, evening sky.)

Jupiter fades to magnitude -2.5. (Apparent diameter 45.0". Cancer, evening sky.)

*February 19th*

21:54 UT - Mercury is at half phase. (50% illuminated, magnitude 0.2, diameter 7.5". Capricornus, pre-dawn sky.)

13:20 UT – The just-past new Moon is north of Neptune. (Neptune: magnitude 8.0, diameter 2.2". Aquarius, not visible.)

*February 21st*

01:17 UT – The waxing crescent Moon is north of Venus. (Venus: 88% illuminated, magnitude -4.0, diameter 11.7". Pisces, evening sky.)

01:40 UT – The waxing crescent Moon is north of Mars. (Mars: 97% illuminated, magnitude 1.3, diameter 4.3". Pisces, evening sky.)

19:53 UT – Venus is 25' south of Mars. (Venus: 88% illuminated, magnitude -4.0, diameter 11.7". Mars: 97% illuminated, magnitude 1.3, diameter 4.3". Pisces, evening sky.)

23:35 UT – The waxing crescent Moon is north of Uranus. (Uranus: magnitude 5.9, diameter 3.4". Pisces, evening sky.)

*February 22nd*

Good opportunity to see Earthshine on the waxing crescent Moon. (Evening sky.)

*February 24th*

16:24 UT – Mercury reaches Greatest Western Elongation. (59% illuminated, magnitude 0.1, diameter 6.9". Capricornus, pre-dawn sky.)

*February 25th*

02:40 UT – The almost first quarter Moon is south of M45, the Pleiades open star cluster. (Taurus, evening sky.)

17:14 UT - First Quarter Moon. (Taurus, evening sky.)

*February 26th*

00:48 UT – The first quarter Moon is north of Aldebaran. (Taurus, evening sky.)

12:15 UT - Neptune is in conjunction with the Sun. (Neptune: magnitude 8.0, diameter 2.2". Aquarius, not visible.)

21:28 UT - Neptune is at its apogee. Distance to Earth: 30.957 AU (Magnitude 8.0, diameter 2.2". Aquarius, not visible.)

# March

*March 1st*

Asteroid 3 Juno fades to magnitude 8.5. (Cancer, evening sky.)

*March 2nd*

05:30 UT – The waxing gibbous Moon is north of asteroid 3 Juno. (Juno: magnitude 8.5. Cancer, evening sky.)

*March 3rd*

03:59 UT – The waxing gibbous Moon is south of Jupiter. (Jupiter: magnitude -2.5, diameter 44.3". Cancer, evening sky.)

*March 4th*

13:00 UT – Mercury is 54' north of asteroid 4 Vesta. (Mercury: 70% illuminated, magnitude 0.0, diameter 6.2". Vesta: magnitude 7.6. Capricornus, pre-dawn sky.)

13:28 UT – The waxing gibbous Moon is south of Regulus (Leo, evening sky.)

16:00 UT - Mercury brightens to magnitude 0.0. (70% illuminated, diameter 6.2". Capricornus, pre-dawn sky.)

18:41 UT – Venus is 5.3' north of Uranus. (Venus: 86% illuminated, magnitude -4.0, diameter 12.2". Uranus: magnitude 5.9, diameter 3.4". Pisces, evening sky.)

Saturn increases its apparent diameter to 17.0". (Magnitude 0.5, Scorpius, pre-dawn sky.)

*March 5th*

18:05 UT - Full Moon. With an apparent diameter of 29.399', this is the smallest full moon of the year and the smallest for the next ten years. The next smallest full moon will occur on January 27th 2032. (Leo, visible all night.)

*March 6th*

20:12 UT - Mercury is at aphelion. Distance to Sun: 0.467 AU (72% illuminated, magnitude 0.0, diameter 6.0", Capricornus, pre-dawn sky.)

*March 8th*

08:24 UT – Asteroid 3 Juno is stationary prior to resuming prograde motion. (Magnitude 8.6. Cancer, evening sky.)

20:26 UT – The waning gibbous Moon is north of Spica. (Virgo, pre-dawn sky.)

*March 11th*

15:57 UT – Mars is 16' south of Uranus. (Mars: 98% illuminated, magnitude 1.3, diameter 4.1". Pisces, evening sky. Uranus: magnitude 5.9, diameter 3.4". Pisces, not visible.)

Mercury leaves Capricornus and enters Aquarius. (77% illuminated, magnitude -0.1, diameter 5.7", pre-dawn sky.)

*March 12th*

09:42 UT – The waning gibbous Moon is north of Saturn (Saturn: magnitude 0.4, diameter 17.2". Scorpius, pre-dawn sky.)

10:28 UT - Saturn is Antares. Separation: 8.4° (Saturn: magnitude 0.4, diameter 17.2". Scorpius, pre-dawn sky.)

15:54 UT – The waning gibbous Moon is north of Antares. (Ophiuchus, pre-dawn sky.)

The Sun leaves Aquarius and enters Pisces.

*March 13th*

17:48 UT - Last Quarter Moon. This is the southernmost last quarter moon of the year. (Ophiuchus, pre-dawn sky.)

*March 14th*

19:12 UT - Saturn is stationary prior to beginning retrograde motion. (Magnitude 0.4, diameter 17.2". Scorpius, pre-dawn sky.)

*March 15th*

10:59 UT – The waning crescent Moon is north of Pluto. (Pluto: magnitude 14.2. Sagittarius, pre-dawn sky.)

*March 16th*

01:29 UT – The waning crescent Moon is north of dwarf planet Ceres. (Ceres: magnitude 8.6. Sagittarius, pre-dawn sky.)

Venus leaves Pisces and enters Aries. (83% illuminated, magnitude -4.0, diameter 12.8", evening sky.)

*March 17th*

23:32 UT – Mercury is 1.6° south of Neptune. (Mercury: 83% illuminated, magnitude -0.3, diameter 5.4"., Aquarius, pre-dawn sky. Neptune: magnitude 8.0, diameter 2.2". Aquarius, not visible.)

Good opportunity to see Earthshine on the waning crescent Moon. (Pre-dawn sky.)

*March 18th*

04:09 UT – The waning crescent Moon is north of asteroid 4 Vesta. (Vesta: magnitude 7.6. Capricornus, pre-dawn sky.)

*March 19th*

01:03 UT – The waning crescent Moon is north of Neptune. (Neptune: magnitude 8.0, diameter 2.2". Aquarius, not visible.)

03:16 UT – The nearly new Moon is north of Mercury. (Mercury: 84% illuminated, magnitude -0.3, diameter 5.4". Aquarius, pre-dawn sky.)

*March 20th*

07:41 UT – First location to see a partial solar eclipse begin.

09:09 UT – First location to see the full solar eclipse begin.

09:36 UT - New Moon. (Pisces, not visible.)

09:45 UT – Total solar eclipse maximum. Visible from north and west Africa, the Arctic, north and east Asia, the Atlantic, Europe and western North America.

10:22 UT – Last location to see the full solar eclipse end.

11:50 UT – Last location to see a partial solar eclipse end.

22:45 UT - Spring Equinox. Spring begins in the northern hemisphere, autumn begins in the southern hemisphere.

*March 21st*

10:09 UT – The waxing crescent Moon is south of Uranus. The Moon may occult Uranus from some parts of the world, but Uranus will be too close to the Sun to be easily found. For observers in the western hemisphere, the occultation will take place during daylight hours and is therefore not visible. (Uranus: magnitude 5.9, diameter 3.3". Pisces, not visible.)

23:15 UT – The waxing crescent Moon is south of Mars. (Mars: 98% illuminated, magnitude 1.3, diameter 4.0". Pisces, evening sky.)

*March 22nd*

21:36 UT – The waxing crescent Moon is south of Venus. (Venus: 81% illuminated, magnitude -4.0, diameter 13.2". Aries, evening sky.)

Mercury brightens to magnitude -0.5. (87% illuminated, diameter 5.3". Aquarius, pre-dawn sky.)

Asteroid 4 Vesta leaves Capricornus and enters Aquarius. (Magnitude 7.6, pre-dawn sky.)

*March 23rd*

Good opportunity to see Earthshine on the waxing crescent Moon. (Evening sky.)

Dwarf planet Ceres brightens to magnitude 8.5. (Sagittarius, pre-dawn sky.)

*March 24th*

08:26 UT – The waxing crescent Moon is south of M45, the Pleiades open star cluster. (Taurus, evening sky.)

Asteroid 3 Juno fades to magnitude 9.0. (Cancer, evening sky.)

*March 25th*

06:09 UT – The waxing crescent Moon is north of Aldebaran. (Taurus, evening sky.)

*March 27th*

07:42 UT - First Quarter Moon. (Gemini, evening sky.)

*March 28th*

Asteroid 2 Pallas leaves Ophiuchus and returns to Hercules. (Magnitude 9.2, pre-dawn sky.)

*March 29th*

Mars leaves Pisces and enters Aries. (99% illuminated, magnitude 1.4, diameter 4.0", evening sky.)

Mercury decreases its apparent diameter to 5.0". (93% illuminated, magnitude -0.8. Aquarius, not visible.)

*March 30th*

09:49 UT – The waxing gibbous Moon is south of Jupiter. (Jupiter: magnitude -2.3, diameter 41.6". Cancer, evening sky.)

Mercury leaves Aquarius and enters Pisces. (94% illuminated, magnitude -0.9, diameter 5.0", not visible.)

Mercury brightens to magnitude -1.0 (94% illuminated, diameter 5.0". Pisces, not visible.)

*March 31st*

20:02 UT – The waxing gibbous Moon is south of Regulus (Leo, evening sky.)

Dwarf planet Ceres leaves Sagittarius and enters Capricornus. (Magnitude 8.5, pre-dawn sky.)

# *April*

### *April 4th*

09:03 UT – Penumbral lunar eclipse begins.

10:17 UT – Partial lunar eclipse begins.

12:01 UT – Total lunar eclipse maximum. Visible from the Antarctica, the Arctic, most of Asia, the Atlantic, Australia, the Indian Ocean, most of north and south America and the Pacific.

12:05 UT - Full Moon. (Virgo, visible all night.)

13:44 UT – Partial lunar eclipse ends.

14:58 UT – Penumbral lunar eclipse ends.

Mercury leaves Pisces and enters Cetus. (98% illuminated, magnitude -1.4, diameter 5.0", not visible.)

### *April 5th*

07:13 UT – The just-past full Moon is north of Spica. (Virgo, visible all night)

21:03 UT - Mercury is at apogee. Distance to Earth: 1.345 AU (99% illuminated, magnitude -1.5, diameter 5.0". Cetus, not visible.)

Mercury leaves Cetus and returns to Pisces. (98% illuminated, magnitude -1.5, diameter 5.0". Pisces, not visible.)

Mercury brightens to magnitude -1.5. (98% illuminated, diameter 5.0". Pisces, not visible.)

*April 6th*

21:55 UT - Uranus is in conjunction with the Sun and is at apogee. Distance to Earth: 20.999 AU. (Magnitude 5.9, diameter 3.3". Pisces, not visible.)

*April 7th*

Venus leaves Aries and enters Taurus. (76% illuminated, magnitude -4.0, diameter 14.3", evening sky.)

*April 8th*

13:31 UT – The waning gibbous Moon is north of Saturn. (Saturn: magnitude 0.3, diameter 17.9", Scorpius, pre-dawn.)

12:44 UT - Mercury is 28' south of Uranus. (Mercury: 100% illuminated, magnitude -1.9, diameter 5.0". Uranus: magnitude 5.9, diameter 3.3". Pisces, not visible.)

16:54 UT - Jupiter is stationary prior to resuming prograde motion. (Magnitude -2.3, diameter 40.6", Cancer, evening sky.)

22:44 UT – The waning gibbous Moon is north of Antares. (Ophiuchus, pre-dawn sky.)

Mercury brightens to magnitude -2.0. (100% illuminated, diameter 5.0". Pisces, not visible.)

*April 10th*

04:00 UT - Mercury is at superior conjunction with the Sun. Distance to Earth: 1.335 AU (100% illuminated, magnitude -2.1, diameter 5.0". Pisces, not visible.)

12:20 UT - Mercury is at maximum brightness. Magnitude: -2.1 (100% illuminated, diameter 5.0". Pisces, not visible.)

Neptune brightens to magnitude 7.9. (Diameter 2.2". Aquarius, pre-dawn sky.)

*April 11th*

00:55 UT - Venus is 2.7° south of M45, the Pleiades open star cluster. (Venus: 75% illuminated, magnitude -4.1, diameter 14.7", Taurus, evening sky.)

17:10 UT – The waning gibbous Moon is north of Pluto. (Pluto: magnitude 14.2. Sagittarius, pre-dawn sky.)

Saturn increases its apparent diameter to 18.0". (Magnitude 0.3, Scorpius, pre-dawn sky.)

*April 12th*

03:44 UT - Last Quarter Moon. (Sagittarius, pre-dawn sky.)

23:32 UT – The just-past last quarter Moon is north of dwarf planet Ceres. (Ceres: magnitude 8.4. Capricornus, pre-dawn sky.)

Jupiter decreases its apparent diameter to 40.0". (Magnitude -2.3, Cancer, evening sky.)

*April 14th*

20:48 UT – Neptune is 2.8° north of asteroid 4 Vesta. (Neptune: magnitude 7.9, diameter 2.2". Vesta: magnitude 7.6. Aquarius, pre-dawn sky.)

Mercury leaves Pisces and enters Aries. (98% illuminated, magnitude -1.8, diameter 5.1", not visible.)

## April 15th

14:50 UT – The waning crescent Moon is north of Neptune. (Neptune: magnitude 7.9, diameter 2.2". Aquarius, pre-dawn sky.)

15:23 UT – The waning crescent Moon is north of asteroid 4 Vesta. (Vesta: magnitude 7.6. Aquarius, pre-dawn sky.)

Good opportunity to see Earthshine on the waning crescent Moon before sunrise. (Pre-dawn sky.)

## April 16th

19:15 UT - Pluto is stationary prior to beginning retrograde motion. (Magnitude 14.2. Sagittarius, pre-dawn sky.)

Venus increases its apparent diameter to 15.0". (74% illuminated, magnitude -4.1. Taurus, evening sky.)

## April 17th

Mercury fades to magnitude -1.5. (95% illuminated, diameter 5.2". Aries, not visible.)

Asteroid 2 Pallas brightens to magnitude 9.0. (Hercules, pre-dawn sky.)

## April 18th

09:18 UT - Venus is at perihelion. Distance to Sun: 0.718 AU. (72% illuminated, magnitude -4.1, diameter 15.3". Taurus, evening sky.)

18:57 UT - New Moon. (Pisces, not visible.)

Asteroid 3 Juno fades to magnitude 9.5. (Cancer, evening sky.)

*April 19th*

09:48 UT – The just-past new Moon is south of Mercury. (Mercury: 91% illuminated, magnitude -1.4, diameter 5.4". Aries, not visible)

19:48 UT - Mercury is at perihelion. Distance to Sun: 0.308 AU (90% illuminated, magnitude -1.4, diameter 5.4". Aries, not visible)

20:40 UT – The just-past new Moon is south of Mars. (Mars: 99% illuminated, magnitude 1.4, diameter 3.9". Aries, not visible.)

Asteroid 2 Pallas is stationary prior to beginning retrograde motion. (Magnitude 9.0. Hercules, pre-dawn sky.)

*April 20th*

21:44 UT – The waxing crescent Moon is south of M45, the Pleiades open star cluster. (Taurus, evening sky.)

*April 21st*

00:01 UT - Venus is Aldebaran. (Venus: 71% illuminated, magnitude -4.1, diameter 15.6", Taurus, evening sky. )

18:24 UT – The waxing crescent Moon is north of Aldebaran. (Taurus, evening sky.)

20:59 UT – The waxing crescent Moon is south of Venus. (Venus: 71% illuminated, magnitude -4.1, diameter 15.6", Taurus, evening sky.)

*April 22nd*

The dust storm season ends on Mars. (99% illuminated, magnitude 1.4, diameter 3.8". Aries, not visible.)

Good opportunity to see Earthshine on the waxing crescent Moon after sunset. (Evening sky.)

The Lyrid meteor shower peaks. Maximum zenith hourly rate: 18. (Moon: waxing crescent. Lyra.)

*April 23rd*

08:05 UT – Mercury is 1.4° north of Mars. (Mercury: 81% illuminated, magnitude -1.1, diameter 5.7". Mars: 99% illuminated, magnitude 1.4, diameter 3.8". Aries, not visible.)

Mercury fades to magnitude -1.0. (80% illuminated, diameter 5.7". Aries, not visible.)

*April 25th*

23:55 UT - First Quarter Moon. (Cancer, evening sky.)

*April 26th*

06:37 UT – The just-past first quarter Moon is south of asteroid 3 Juno. (Juno: magnitude 9.6. Cancer, evening sky.)

17:42 UT – The just-past first quarter Moon is south of Jupiter. (Jupiter: magnitude -2.2, diameter 38.4", Cancer, evening sky.)

*April 27th*

Mercury leaves Aries and enters Taurus. (67% illuminated, magnitude -0.7, diameter 6.2", evening sky.)

*April 28th*

04:55 UT – The waxing gibbous Moon is south of Regulus. (Leo, evening sky.)

*April 29th*

Mercury fades to magnitude -0.5. (62% illuminated, diameter 6.5". Taurus, evening sky.)

*April 30th*

17:39 UT - Mercury is 1.7° south of the Pleiades. (Mercury: 57% illuminated, magnitude -0.4, diameter 6.7", Taurus, evening sky.)

# *May*

## *May 2nd*

12:02 UT – The waxing gibbous Moon is north of Spica. (Virgo, evening sky.)

19:42 UT - Mercury reaches half phase. (50% illuminated, magnitude -0.2, diameter 7.1", Taurus, evening sky.)

Mars leaves Aries and enters Taurus. (100% illuminated, magnitude 1.4, diameter 3.8", evening sky.)

## *May 4th*

02:00 UT - Mercury fades to magnitude 0.0. (46% illuminated, diameter 7.3", Taurus, evening sky.)

03:42 UT - Full Moon. (Libra, visible all night.)

## *May 5th*

16:07 UT – The just-past full Moon is north of Saturn. (Saturn: magnitude 0.1, diameter 18.4", Scorpius, visible all night.)

The Eta Aquariid meteor shower peaks. Maximum zenith hourly rate: 65. (Moon: just past full. Aquarius.)

## *May 6th*

03:48 UT – The waning gibbous Moon is north of Antares. (Ophiuchus, pre-dawn sky.)

Mars fades to magnitude 1.5. (100% illuminated, diameter 3.8". Taurus, evening sky.)

## *May 7th*

04:48 UT – Mercury reaches Greatest Eastern Elongation. (37% illuminated, magnitude 0.4, diameter 7.9", Taurus, evening sky.)

Mercury fades to magnitude 0.5. (38% illuminated, diameter 7.9". Taurus, evening sky.)

## *May 8th*

20:29 UT – The waning gibbous Moon is north of Pluto. (Pluto: magnitude 14.1. Sagittarius, pre-dawn sky.)

Venus leaves Taurus and enters Gemini. (64% illuminated, magnitude -4.1, diameter 17.7", evening sky.)

## *May 10th*

17:12 UT – The almost last quarter Moon is north of dwarf planet Ceres. (Ceres: magnitude 8.1. Capricornus, pre-dawn sky.)

17:42 UT – Mars is 3.9° south of M45, the Pleiades open star cluster. (Mars: 100% illuminated, magnitude 1.5, diameter 3.8". Taurus, evening sky.)

## *May 11th*

10:36 UT - Last Quarter Moon. (Capricornus, pre-dawn sky.)

17:47 UT – Mercury is 7.9° north of Aldebaran. (Mercury: 26% illuminated, magnitude 1.0, diameter 8.9". Taurus, evening sky.)

Mercury fades to magnitude 1.0 (26% illuminated, diameter 8.9". Taurus, evening sky.)

Jupiter fades to magnitude -2.0. (Magnitude -2.1, diameter 36.8", Cancer, evening sky.)

*May 12th*

20:44 UT – The just-past last quarter Moon is north of Neptune. (Neptune: magnitude 7.9, diameter 2.2". Aquarius, pre-dawn sky.)

*May 13th*

18:26 UT – The waning crescent Moon is north of asteroid 4 Vesta. (Vesta: magnitude 7.5. Aquarius, pre-dawn sky.)

*May 14th*

The Sun leaves Aries and enters Taurus.

*May 15th*

13:25 UT – The waning crescent Moon is north of Uranus. This is a very close conjunction and in some parts of the world, an occultation may be visible. (Uranus: magnitude 5.9, diameter 3.4". Pisces, pre-dawn sky.)

Saturn returns to Libra from Scorpius. (Magnitude 0.1, diameter 17.4", pre-dawn sky.)

Good opportunity to see Earthshine on the waning crescent Moon before sunrise.

*May 16th*

Mercury increases its apparent diameter to 10.0". (16% illuminated, magnitude 1.8. Taurus, Evening sky.)

Saturn brightens to magnitude 0.0. (Diameter 18.4", Libra, pre-dawn sky.)

Dwarf planet Ceres brightens to magnitude 8.0. (Capricornus, pre-dawn sky.)


*May 18th*

04:13 UT - New Moon. (Taurus, not visible.)

04:37 UT – The new Moon is south of M45, the Pleiades open star cluster. (Taurus, not visible.)

18:40 UT – The new Moon is south of Mars. (Mars: 100% illuminated, magnitude 1.5, diameter 3.7". Taurus, not visible.)


*May 19th*

01:59 UT – The just-past new Moon is north of Aldebaran. (Taurus, not visible)

05:20 UT – The just-past new Moon is south of Mercury. (Mercury: 10% illuminated, magnitude 2.4, diameter 10.6". Taurus, not visible)

10:35 UT - Mercury is stationary prior to beginning retrograde motion. (10% illuminated, magnitude 2.5, diameter 10.7". Taurus, not visible.)


*May 20th*

Asteroid 3 Juno fades to magnitude 10.0. (Cancer, evening sky.)

*May 21st*

20:52 UT – The waxing crescent Moon is south of Venus. (Venus: 58% illuminated, magnitude -4.2, diameter 19.8", Gemini, evening sky.)

Good opportunity to see Earthshine on the waxing crescent Moon after sunset.

*May 22nd*

Venus increases its apparent diameter to 20.0". (58% illuminated, magnitude -4.2. Gemini, evening sky.)

*May 23rd*

02:00 UT - Saturn is at opposition. (Magnitude 0.0, diameter 18.5", Libra, visible all night.)

03:00 UT - Saturn is at perigee. Distance to Earth: 8.967 AU (Magnitude 0.0, diameter 18.5", Libra, visible all night.)

11:27 UT - Saturn is at its maximum brightness for 2015. Magnitude: 0.0 (Diameter 18.5", Libra, visible all night.)

17:41 UT – Jupiter is 3.6° north of asteroid 3 Juno. (Jupiter: magnitude -2.0, diameter 35.4". Juno: magnitude 10.0. Cancer, evening sky.)

*May 24th*

06:31 UT – The waxing crescent Moon is south of Jupiter. (Jupiter: magnitude -2.0, diameter 35.3", Cancer, evening sky.)

06:39 UT – The waxing crescent Moon is south of asteroid 3 Juno. (Juno: magnitude 10.0. Cancer, evening sky.)

*May 25th*

09:41 UT – The almost first quarter Moon is south of Regulus. (Leo, evening sky.)

17:18 UT - First Quarter Moon. (Leo, evening sky.)


*May 27th*

06:41 UT - Mars is 6° north of Aldebaran. (Mars: 100% illuminated, magnitude 1.5, diameter 3.7". Taurus, not visible.)

15:05 UT – Mercury is 1.6° south of Mars. (Mercury: 1% illuminated, magnitude 4.9, diameter 12.0". Mars: 100% illuminated, magnitude 1.5, diameter 3.7". Taurus, not visible.)

Jupiter decreases its apparent diameter to 35.0". (Magnitude -2.0, Cancer, evening sky.)


*May 29th*

Asteroid 4 Vesta leaves Aquarius and enters Pisces. (Magnitude 7.4, pre-dawn sky.)


*May 30th*

16:54 UT - Mercury is at inferior conjunction with the Sun. Distance to Earth: 0.549 AU. (0% illuminated, magnitude 5.7, diameter 12.2". Taurus, not visible.)

20:52 UT – Venus is 4.1° south of Pollux. (Venus: 54% illuminated, magnitude -4.3, diameter 21.7", Gemini, evening sky.)

Asteroid 3 Juno leaves Cancer and enters Leo. (Magnitude 10.1, evening sky.)

*May 31st*

03:24 UT - Mercury is at perigee. Distance to Earth: 0.549 AU. (0% illuminated, magnitude 5.6, diameter 12.2". Taurus, not visible.)

# *June*

### *June 1st*

08:24 UT – Asteroid 2 Pallas reaches its maximum brightness. Magnitude: 8.9 (Hercules, pre-dawn sky.)

19:18 UT – The almost full Moon is north of Saturn. (Saturn: magnitude 0.1, diameter 18.4", Libra, visible all night.)

### *June 2nd*

10:38 UT – The almost full Moon is north of Antares. (Ophiuchus, visible all night.)

16:19 UT - Full Moon. (Ophiuchus, visible all night.)

19:24 UT - Mercury is at aphelion. Distance to Sun: 0.467 AU (1% illuminated, magnitude 4.8, diameter 12.1". Taurus, not visible)

### *June 3rd*

Venus leaves Gemini and enters Cancer. (52% illuminated, magnitude -4.3, diameter 22.7", evening sky.)

Asteroid 2 Pallas is at perigee. Distance to Earth: 2.399 AU. (Magnitude 8.9. Hercules, pre-dawn sky.)

### *June 5th*

04:12 UT – The waning gibbous Moon is north of Pluto. (Pluto: magnitude 14.1. Sagittarius, pre-dawn sky.)

*June 6th*

01:03 UT - Dwarf planet Ceres is stationary prior to beginning retrograde motion. (Magnitude 7.8. Capricornus, pre-dawn sky.)

09:06 UT - Venus is at half phase. (50% illuminated, magnitude -4.3, diameter 23.4", Cancer, evening sky.)

18:30 UT - Venus is at Greatest Eastern Elongation. (50% illuminated, magnitude -4.3, diameter 23.4", Cancer, evening sky.)

*June 7th*

00:57 UT – The waning gibbous Moon is north of dwarf planet Ceres. (Ceres: magnitude 7.8. Capricornus, pre-dawn sky.)

The Arietid meteor shower peaks. Maximum zenith hourly rate: 54. (Moon: waning gibbous. Aries.)

*June 8th*

Jupiter returns to Leo from Cancer. (Magnitude -1.9, diameter 34.0", evening sky.)

*June 9th*

01:32 UT – The almost last-quarter Moon is north of Neptune. (Neptune: magnitude 7.9, diameter 2.2". Aquarius, pre-dawn sky.)

15:41 UT - Last Quarter Moon. With an apparent diameter of 32.304', this is the largest last quarter Moon of the year and the largest for 10 years. The most recent larger last quarter Moon was on July 16th, 1998. (Aquarius, pre-dawn sky.)

*June 10th*

17:59 UT – The waning crescent Moon is north of asteroid 4 Vesta. (Vesta: magnitude 7.3. Pisces, pre-dawn sky.)

*June 11th*

19:39 UT - Mercury is stationary prior to resuming prograde motion. (11% illuminated, magnitude 2.4, diameter 10.8". Taurus, not visible.)

20:20 UT – The waning crescent Moon is south of Uranus. (Uranus: magnitude 5.9, diameter 3.4". Pisces, pre-dawn sky.)

Venus increases its apparent diameter to 25.0". (47% illuminated, magnitude -4.3. Cancer, evening sky.)

*June 12th*

15:59 UT - Neptune is stationary prior to beginning retrograde motion. (Magnitude 7.9, diameter 2.3". Aquarius, pre-dawn sky.)

*June 13th*

03:13 UT – Venus is 0.9° north of M44, the Praesepe open star cluster. (Venus: 46% illuminated, magnitude -4.3, diameter 25.5". Cancer, evening sky.)

Good opportunity to see Earthshine on the waning crescent Moon before sunrise.

*June 14th*

13:12 UT – Asteroid 2 Pallas is at opposition. (Magnitude 8.9. Hercules, visible all night.)

15:56 UT - Mars is in conjunction with the Sun. Distance to Earth: 2.568 AU (100% illuminated, magnitude 1.5, diameter 3.6". Taurus, not visible.)

17:16 UT – The waning crescent Moon is south of M45, the Pleiades open star cluster. (Taurus, not visible.)

*June 15th*

02:20 UT – The waning crescent Moon is south of Mercury. (Mercury: 17% illuminated, magnitude 1.8, diameter 10.1". Taurus, not visible.)

12:17 UT – The waning crescent Moon is north of Aldebaran. (Taurus, not visible.)

Mercury decreases its apparent diameter to 10.0". (17% illuminated, magnitude 1.8. Taurus, pre-dawn sky.)

*June 16th*

14:05 UT - New Moon. (Taurus, not visible.)

14:20 UT – The just-past new Moon is south of Mars. (Mars: 100% illuminated, magnitude 1.5, diameter 3.6". Taurus, not visible.)

*June 18th*

12:41 UT - Spring begins in the northern hemisphere on Mars. (Mars: 100% illuminated, magnitude 1.5, diameter 3.6". Taurus, not visible.)

*June 20th*

09:32 UT – The waxing crescent Moon is south of Venus. (Venus: 41% illuminated, magnitude -4.4, diameter 27.9", Cancer, evening sky.)

Mercury brightens to magnitude 1.0 (26% illuminated, diameter 9.0". Taurus, pre-dawn sky.)

Good opportunity to see Earthshine on the waxing crescent Moon after sunset.

*June 21st*

00:50 UT – The waxing crescent Moon is south of Jupiter. (Jupiter: magnitude -1.8, diameter 33.0", Leo, evening sky.)

09:19 UT – The waxing crescent Moon is south of asteroid 3 Juno (Juno: magnitude 10.3. Leo, evening sky.)

16:38 UT - Northern Solstice. Summer begins in the northern hemisphere, winter begins in the southern hemisphere.

21:00 UT – The waxing crescent Moon is south of Regulus. (Leo, evening sky.)

The Sun leaves Taurus and enters Gemini.

Asteroid 4 Vesta leaves Pisces and enters Cetus. (Magnitude 7.2, pre-dawn sky.)

*June 22nd*

Uranus increases its apparent diameter to 3.5". (Magnitude 5.9, Pisces, pre-dawn sky.)

*June 23rd*

Uranus brightens to magnitude 5.8. (Apparent diameter 3.5". Pisces, pre-dawn sky.)

*June 24th*

03:07 UT – Mercury is 2.0° north of Aldebaran (Mercury: 35% illuminated, magnitude 0.6, diameter 8.2". Taurus, pre-dawn sky.)

11:02 UT - First Quarter Moon. With an apparent diameter of 29.585', this is the smallest first quarter moon of the year. (Virgo, evening sky.)

17:06 UT - Mercury is at Greatest Western Elongation from the Sun. (37% illuminated, magnitude 0.5, diameter 8.1". Taurus, pre-dawn sky.)

Mercury brightens to magnitude 0.5. (35% illuminated, diameter 8.2". Taurus, pre-dawn sky.)

Mars leaves Taurus and enters Gemini. (100% illuminated, magnitude 1.5, diameter 3.5", not visible.)

Dwarf planet Ceres brightens to magnitude 7.5. (Capricornus, pre-dawn sky.)


*June 25th*

Venus leaves Cancer and enters Leo. (38% illuminated, magnitude -4.4, diameter 29.8", evening sky.)

Venus increases its apparent diameter to 30.0". (38% illuminated, magnitude -4.4. Leo, evening sky.)

Asteroid 2 Pallas fades to magnitude 9.0. (Hercules, evening sky.)


*June 26th*

04:14 UT – The waxing gibbous Moon is north of Spica. (Virgo, evening sky.)

*June 29th*

02:54 UT – The waxing gibbous Moon is north of Saturn. (Saturn: magnitude 0.2, diameter 18.1", Libra, evening sky.)

Dwarf planet Ceres leaves Capricornus and enters Microscopium. (Magnitude 7.5, pre-dawn sky.)

Saturn decreases its apparent diameter to 18.0". (Magnitude 0.2, Libra, evening sky.)

*June 30th*

00:00 UT - Mercury brightens to magnitude 0.0. (49% illuminated, diameter 7.1". Taurus, pre-dawn sky, pre-dawn sky.)

04:54 UT - Mercury is at half phase. (50% illuminated, magnitude 0.0, diameter 7.1". Taurus, pre-dawn sky, pre-dawn.)

# *July*

### *July 1st*

03:49 UT – Venus is 20' south of Jupiter. (Venus: 34% illuminated, magnitude -4.4, diameter 32.5". Jupiter: magnitude -1.8, diameter 32.4", Leo, evening sky.)

### *July 2nd*

02:19 UT – Full Moon. The first of two July Full Moons occurs – the second occurs on the 31st. This is also the southernmost full moon of the year. (Sagittarius, visible all night.)

02:58 UT – Asteroid 3 Juno is 1.1° south of Regulus. (Juno: magnitude 10.4. Leo, evening sky.)

07:11 UT - Pluto has reached its maximum brightness for 2015. Magnitude: 14.1. (Sagittarius, visible all night.)

10:45 UT – The just-past full Moon is north of Pluto. (Pluto: magnitude 14.1. Sagittarius, visible all night.)

Venus brightens to magnitude -4.5. (33% illuminated, diameter 32.8". Leo, evening sky.)

### *July 4th*

06:33 UT – The waning gibbous Moon is north of dwarf planet Ceres. (Ceres: magnitude 7.4. Microscopium, pre-dawn sky.)

08:03 UT - Pluto is at perigee. Distance to Earth: 31.887 AU. (Pluto: magnitude 14.1. Sagittarius, visible all night.)

Mercury brightens to magnitude -0.5. (60% illuminated, diameter 6.5". Taurus, pre-dawn sky.)

## *July 6th*

07:05 UT - Pluto is at opposition. Distance to Earth: 31.888 AU. (Magnitude 14.1. Sagittarius, visible all night.)

11:10 UT – The waning gibbous Moon is north of Neptune. (Neptune: magnitude 7.9, diameter 2.3". Aquarius, pre-dawn sky.)

13:12 UT – The Earth is at aphelion. Distance to Sun: 1.017 AU.

Venus increases its apparent diameter to 35.0". (30% illuminated, magnitude -4.5. Leo, evening sky.)

Asteroid 4 Vesta brightens to magnitude 7.0. (Cetus, pre-dawn sky.)

## *July 7th*

Mercury leaves Taurus and enters Orion. (70% illuminated, magnitude -0.7, diameter 6.0", pre-dawn sky.)

Asteroid 3 Juno fades to magnitude 10.5. (Leo, evening sky.)

## *July 8th*

14:09 UT – The almost last quarter Moon is north of asteroid 4 Vesta. (Vesta: magnitude 7.0. Cetus, pre-dawn sky.)

20:24 UT - Last Quarter Moon. (Pisces, pre-dawn sky.)

Mercury leaves Orion and enters Gemini. (73% illuminated, magnitude -0.8, diameter 5.9", pre-dawn sky.)

*July 9th*

01:07 UT – The just-past last quarter Moon is south of Uranus. (Uranus: magnitude 5.8, diameter 3.5". Pisces, pre-dawn sky.)

Neptune brightens to magnitude 7.8. (Diameter 2.3". Aquarius, pre-dawn sky.)

*July 10th*

Mercury brightens to magnitude -1.0 (78% illuminated, diameter 5.8". Gemini, pre-dawn sky.)

*July 11th*

12:33 UT - Mars is at apogee. Distance to Earth: 2.587 AU (100% illuminated, magnitude 1.6, diameter 3.6". Gemini, not visible.)

21:44 UT – The waning crescent Moon is south of M45, the Pleiades open star cluster. (Taurus, pre-dawn sky.)

*July 12th*

05:24 UT - Venus reaches its maximum brightness. Magnitude: -4.47. (Illuminated 25%, diameter 38.6", Leo, evening sky.)

18:55 UT – The waning crescent Moon is north of Aldebaran. (Taurus, pre-dawn sky.)

*July 13th*

Good opportunity to see Earthshine on the waning crescent Moon before sunrise.

*July 14th*

Venus increases its apparent diameter to 40.0". (23% illuminated, magnitude -4.5. Leo, evening sky.)

*July 15th*

05:21 UT – The almost new Moon is south of Mercury. (Mercury: 91% illuminated, magnitude -1.4, diameter 5.3". Gemini, not visible.)

06:05 UT - Venus is Regulus. (Venus: 22% illuminated, magnitude -4.5, diameter 40.5", Leo, evening sky.)

08:26 UT – The almost new Moon is south of Mars. (Mars: 100% illuminated, magnitude 1.6, diameter 3.6". Gemini, not visible.)

Mercury brightens to magnitude -1.5. (91% illuminated, diameter 5.3". Gemini, not visible.)

*July 16th*

01:24 UT - New Moon. (Gemini, not visible.)

05:24 UT – Mercury is 6' south of Mars. (Mercury: 93% illuminated, magnitude -1.5, diameter 5.3". Mars: 100% illuminated, magnitude 1.6, diameter 3.6". Gemini, not visible.)

19:06 UT - Mercury is at perihelion. Distance to Sun: 0.308 AU (94% illuminated, magnitude -1.5, diameter 5.2". Gemini, not visible.)

*July 18th*

19:33 UT – The waxing crescent Moon is south of Jupiter. (Jupiter: magnitude -1.7, diameter 31.6", Leo, evening sky.)

*July 19th*

01:31 UT – The waxing crescent Moon is south of Venus. (Venus: 18% illuminated, magnitude -4.5, diameter 43.2", Leo, evening sky.)

02:40 UT – The waxing crescent Moon is south of Regulus. (Leo, evening sky.)

02:58 UT – The waxing crescent Moon is south of asteroid 3 Juno. (Juno: magnitude 10.5. Leo, evening sky.)

Good opportunity to see Earthshine on the waxing crescent Moon after sunset.

*July 20th*

19:33 UT – Mercury is 5.4° south of Pollux (Mercury: 99% illuminated, magnitude -1.9, diameter 5.1". Gemini, not visible.)

*July 21st*

The Sun leaves Gemini and enters Cancer.

Mercury brightens to magnitude -2.0 and decreases its apparent diameter to 5.0". (99% illuminated. Gemini, not visible.)

Venus increases its apparent diameter to 45.0". (17% illuminated, magnitude -4.4. Leo, evening sky.)

*July 22nd*

11:02 UT - Venus is stationary prior to beginning retrograde motion. (16% illuminated, magnitude -4.4, diameter 45.3", Leo, evening sky.)

Mercury leaves Gemini and enters Cancer. (100% illuminated, magnitude -2.0, diameter 5.0", not visible.)

*July 23rd*

04:57 UT – Dwarf planet Ceres is at perigee. Distance to Earth: 1.939 AU (Ceres: magnitude 7.3. Microscopium, visible all night.)

09:08 UT - The waxing crescent Moon is north of Spica. (Virgo, evening sky)

09:16 UT - Mercury reaches its maximum brightness. Magnitude: -2.1 (99% illuminated, diameter 5.0". Cancer, not visible)

19:24 UT - Mercury is at superior conjunction with the Sun. Distance to Earth: 1.337 AU (100% illuminated, magnitude -2.1, diameter 5.0". Cancer, not visible)


*July 24th*

04:04 UT - First Quarter Moon. (Virgo, evening sky.)

09:51 UT - Dwarf planet Ceres reaches its maximum brightness. Magnitude: 7.3 (Microscopium, visible all night.)

Mercury fades to magnitude -2.0 (99% illuminated, diameter 5.0". Cancer, not visible.)


*July 25th*

Dwarf planet Ceres leaves Microscopium and returns to Sagittarius. (Magnitude 7.3. All night.)


*July 26th*

08:04 UT – The waxing gibbous Moon is north of Saturn. (Saturn: magnitude 0.4, diameter 17.4", Libra, evening sky.)

13:09 UT - Uranus is stationary prior to beginning retrograde motion. (Magnitude 5.8, diameter 3.5". Pisces, pre-dawn sky.)

21:37 UT - Mercury is 0.5° north of M44, the Praesepe open star cluster. (Mercury: 99% illuminated, magnitude -1.7, diameter 5.0". Cancer, not visible.)

22:22 UT - Mercury is at apogee. Distance to Earth: 1.342 AU (99% illuminated, magnitude -1.7, diameter 5.0". Cancer, not visible.)

*July 27th*

05:10 UT – The waxing gibbous Moon is north of Antares. (Ophiuchus, evening sky.)

09:19 UT - Dwarf planet Ceres is at opposition. (Magnitude 7.3. Sagittarius, visible all night.)

*July 28th*

Mercury fades to magnitude -1.5. (98% illuminated, diameter 5.0". Cancer, not visible.)

Venus increases its apparent diameter to 50.0". (11% illuminated, magnitude -4.4. Leo, evening sky.)

*July 29th*

The Southern Delta Aquariid meteor shower peaks. Maximum zenith hourly rate: 16. (Moon: waxing gibbous. Aquarius.)

The Beta Cassiopeid meteor shower peaks. Maximum zenith hourly rate: 10. (Moon: waxing gibbous. Cassiopeia.)

*July 31st*

03:50 UT – Mars is 5.8° south of Pollux. (Mars: 100% illuminated, magnitude 1.7, diameter 3.6". Gemini, not visible.)

06:18 UT – The almost full Moon is north of dwarf planet Ceres. (Ceres: magnitude 7.3. Sagittarius, visible all night.)

10:43 UT – Full Moon. The second of two July Full Moons (mistakenly called a "blue Moon"). The first occurred on the 2nd. (Capricornus, visible all night.)

# *August*

### *August 1st*

Mercury leaves Cancer and enters Leo. (95% illuminated, magnitude -1.1, diameter 5.0", not visible.)

### *August 2nd*

06:00 UT - Saturn is stationary prior to resuming prograde motion. (Magnitude 0.4, diameter 17.2", Libra, evening sky.)

14:55 UT – The waning gibbous Moon is north of Neptune. (Neptune: magnitude 7.8, diameter 2.3". Aquarius, pre-dawn sky.)

Mercury fades to magnitude -1.0. (94% illuminated, diameter 5.1". Leo, not visible.)

### *August 3rd*

Saturn fades to magnitude 0.5. (Apparent diameter 17.2", Libra, evening sky.)

### *August 5th*

01:14 UT – The waning gibbous Moon is north of asteroid 4 Vesta. (Vesta: magnitude 6.7. Cetus, pre-dawn sky.)

09:49 UT – Mercury is 8.2° north of Venus. (Mercury: 90% illuminated, magnitude -0.8, diameter 5.1". Venus: 4% illuminated, magnitude -4.2, diameter 54.8". Leo, not visible.)

11:29 UT – The waning gibbous Moon is south of Uranus. (Uranus: magnitude 5.8, diameter 3.6". Pisces, pre-dawn sky.)

Mars leaves Gemini and enters Cancer. (99% illuminated, magnitude 1.7, diameter 3.6", not visible.)

*August 7th*

02:02 UT - Last Quarter Moon. (Aries, pre-dawn sky.)

07:34 UT - Mercury is 32' north of Jupiter. (Mercury: 88% illuminated, magnitude -0.6, diameter 5.2". Jupiter: magnitude -1.7, diameter 31.0". Leo, evening sky.)

20:23 UT - Mercury is 53' north of Regulus. (88% illuminated, magnitude -0.6, diameter 5.2". Leo, evening sky.)

Saturn decreases its apparent diameter to 17.0". (Magnitude 0.5, Libra, evening sky.)

*August 8th*

01:25 UT – The waning crescent Moon is south of M45, the Pleiades open star cluster. (Taurus, pre-dawn sky.)

03:25 UT – Asteroid 2 Pallas is stationary prior to resuming prograde motion. (Magnitude 9.3. Hercules, evening sky.)

19:42 UT - Venus is at aphelion. Distance to Sun: 0.728 AU (3% illuminated, magnitude -4.1, diameter 56.2". Leo, not visible.)

22:12 UT – The waning crescent Moon is north of Aldebaran. (Taurus, pre-dawn sky.)

Mercury fades to magnitude -0.5. (88% illuminated, diameter 5.2". Leo, evening sky.)

*August 9th*

22:52 UT - Jupiter is 0.4° north of Regulus. (Jupiter: magnitude -1.7, diameter 30.9". Leo, not visible.)

Venus fades to magnitude -4.0 (2% illuminated, diameter 56.6". Leo, not visible.)

*August 11th*

The Sun leaves Cancer and enters Leo.

Good opportunity to see Earthshine on the waning crescent Moon before sunrise. (Pre-dawn sky.)

*August 12th*

Asteroid 4 Vesta brightens to magnitude 6.5. (Cetus, pre-dawn sky.)

The Perseid meteor shower peaks. Maximum zenith hourly rate: 100. (Moon: waning crescent. Perseus.)

*August 13th*

02:55 UT – The almost new Moon is south of Mars. (Mars: 99% illuminated, magnitude 1.7, diameter 3.7", Cancer, pre-dawn sky.)

Dwarf planet Ceres fades to magnitude 7.5. (Sagittarius, evening sky.)

*August 14th*

14:53 UT - New Moon. (Leo, not visible.)

18:20 UT – The just-past new Moon is north of Venus. (Venus: 1% illuminated, magnitude -3.9, diameter 57.8". Leo, not visible.)

*August 15th*

02:14 UT – Asteroid 4 Vesta is stationary prior to beginning retrograde motion. (Magnitude 6.5. Cetus, pre-dawn sky.)

09:00 UT – The just-past new Moon is south of Regulus. (Leo, not visible.)

12:36 UT – The just-past new Moon is south of Jupiter. (Jupiter: magnitude -1.7, diameter 30.8". Leo, not visible.)

19:24 UT - Venus is at inferior conjunction with the Sun. Distance to Earth: 0.288 AU. (1% illuminated, magnitude -3.9, diameter 57.8". Leo, not visible.)

*August 16th*

00:30 UT - Venus is at perigee. Distance to Earth: 0.288 AU. (1% illuminated, magnitude -4.0, diameter 57.8". Leo, not visible.)

15:55 UT – The waxing crescent Moon is south of Mercury. (Mercury: 78% illuminated, magnitude -0.2, diameter 5.6". Leo, evening sky.)

22:31 UT – The waxing crescent Moon is south of asteroid 3 Juno (Juno: magnitude 10.7. Leo, not visible.)

Venus brightens to magnitude -4.0 (1% illuminated, magnitude -4.0, diameter 57.8". Leo, not visible.)

*August 18th*

Good opportunity to see Earthshine on the waxing crescent Moon after sunset. (Evening sky.)

*August 19th*

16:15 UT – Mercury is 1.3° south of asteroid 3 Juno. (Mercury: 75% illuminated, magnitude -0.1, diameter 5.8". Leo, evening sky. Juno: magnitude 10.7. Leo, not visible.)

19:22 UT - The waxing crescent Moon is north of Spica. (Virgo, evening sky.)

Venus returns to Cancer from Leo. (2% illuminated, magnitude -4.0, diameter 57.5", not visible.)

Uranus brightens to magnitude 5.7. (Apparent diameter 3.6". Pisces, pre-dawn sky.)

*August 20th*

07:02 UT – Mars is through M44, the Praesepe open star cluster. (Mars: 99% illuminated, magnitude 1.8, diameter 3.7". Cancer, pre-dawn sky.)

*August 21st*

Asteroid 2 Pallas fades to magnitude 9.5. (Hercules, evening sky.)

*August 22nd*

18:39 UT – The first quarter Moon is north of Saturn. (Saturn: magnitude 0.5, diameter 16.6", Libra, evening sky.)

19:31 UT - First Quarter Moon. (Libra, evening sky.)

## *August 23rd*

03:00 UT - Mercury fades to magnitude 0.0. (71% illuminated, diameter 6.0". Leo, evening sky.)

18:42 UT - The just-past first quarter Moon is north of Antares. (Ophiuchus, evening sky.)

Mercury leaves Leo and enters Virgo. (71% illuminated, magnitude 0.0, diameter 6.0", evening sky.)

## *August 26th*

05:05 UT - The waxing gibbous Moon is north of Pluto. (Pluto: magnitude 14.1. Sagittarius, evening sky.)

22:02 UT - Jupiter is in conjunction with the Sun. Distance to Earth: 6.399 AU (Jupiter: magnitude -1.7, diameter 30.8". Leo, not visible.)

## *August 27th*

00:13 UT - Jupiter is at apogee. Distance to Earth: 6.399 AU. (Magnitude -1.7, diameter 30.8". Leo, not visible.)

07:33 UT – The waxing gibbous Moon is north of dwarf planet Ceres. (Ceres: magnitude 7.6. Sagittarius, evening sky.)

## *August 29th*

05:16 UT - Venus is 9.4° south of Mars. (Venus: 7% illuminated, magnitude -4.3, diameter 53.2". Mars: 99% illuminated, magnitude 1.8, diameter 3.7". Cancer, pre-dawn sky.)

18:35 UT - Full Moon. (Aquarius, visible all night.)

18:42 UT - Mercury is at aphelion. Distance to Sun: 0.4667 AU (63% illuminated, magnitude 0.1, diameter 6.5". Virgo, evening sky.)

*August 30th*

00:004 UT – The just-past full Moon is north of Neptune. (Neptune: magnitude 7.8, diameter 2.3". Aquarius, visible all night.)

*August 31st*

10:26 UT - Neptune is at perigee. Distance to Earth: 28.953 AU (Magnitude 7.8, diameter 2.3". Aquarius, visible all night.)

12:31 UT - Neptune reaches its maximum brightness. Magnitude: 7.8 (Diameter: 2.3". Aquarius, visible all night.)

Asteroid 3 Juno leaves Leo and enters Virgo. (Magnitude 10.7, not visible.)

# *September*

## *September 1st*

11:05 UT – The waning gibbous Moon is north of asteroid 4 Vesta. (Vesta: magnitude 6.3. Cetus, pre-dawn sky.)

11:41 UT - Neptune is at opposition. Distance to Earth: 28.953 AU. (Magnitude 7.8, diameter 2.3". Aquarius, visible all night.)

15:35 UT – The waning gibbous Moon is south of Uranus. (Uranus: magnitude 5.7, diameter 3.6". Pisces, pre-dawn sky.)

## *September 2nd*

17:50 UT - Venus is 9.4° south of Mars. (Venus: 10% illuminated, magnitude -4.4, diameter 50.6". Mars: 98% illuminated, magnitude 1.8, diameter 3.7", Cancer, pre-dawn sky.)

## *September 3rd*

Venus decreases its apparent diameter to 50.0". (11% illuminated, magnitude -4.4. Cancer, pre-dawn sky.)

## *September 4th*

10:18 UT – Mercury is at Greatest Eastern Elongation. (55% illuminated, magnitude 0.2, diameter 7.1". Virgo, evening sky.)

10:58 UT – The waning gibbous Moon is south of M45, the Pleiades open star cluster. (Taurus, pre-dawn sky.)

18:11 UT - Venus is stationary prior to resuming prograde motion. (12% illuminated, magnitude -4.4, diameter 49.2", Cancer, pre-dawn sky.)

*September 5th*

04:48 UT – The almost last quarter gibbous Moon is north of Aldebaran. (Taurus, pre-dawn sky.)

09:54 UT - Last Quarter Moon. (Taurus, pre-dawn sky.)

Venus brightens to magnitude -4.5. (13% illuminated, diameter 48.8". Cancer, pre-dawn sky.)

Mars leaves Cancer and enters Leo. (98% illuminated, magnitude 1.8, diameter 3.8", pre-dawn sky.)

*September 7th*

08:54 UT - Mercury is at half phase. (50% illuminated, magnitude 0.3, diameter 7.4". Virgo, evening sky.)

*September 10th*

05:18 UT – The waning crescent Moon is north of Venus. (18% illuminated, magnitude -4.5, diameter 45.0", Cancer, pre-dawn sky.)

22:52 UT – The waning crescent Moon is south of Mars. (Mars: 98% illuminated, magnitude 1.8, diameter 3.8", Leo, pre-dawn sky.)

Good opportunity to see Earthshine on the waning crescent Moon before sunrise. (Pre-dawn sky.)

Venus decreases its apparent diameter to 45.0". (17% illuminated, magnitude -4.5. Cancer, pre-dawn sky.)

*September 11th*

17:45 UT – The waning crescent Moon is south of Regulus. (Leo, pre-dawn sky.)

Mercury fades to magnitude 0.5. (43% illuminated, diameter 7.9". Virgo, evening sky.)

Asteroid 2 Pallas leaves Hercules and returns to Ophiuchus. (Magnitude 9.6, evening sky.)

*September 12th*

05:36 UT – The almost new Moon is south of Jupiter. (Jupiter: magnitude -1.7, diameter 30.9". Leo, not visible.)

*September 13th*

04:42 UT – First location to see the partial solar eclipse begin.

06:41 UT - New Moon. This is the farthest new moon of the year. (Leo, not visible.)

06:55 UT – Partial solar eclipse maximum. Visible from southern Africa, Antarctica, the Atlantic and the Indian Ocean.

09:07 UT – Last location to see the partial solar eclipse end.

Saturn decreases its apparent diameter to 16.0". (Magnitude 0.6, Libra, evening sky.)

*September 14th*

01:07 UT – The just-past new Moon is south of asteroid 3 Juno (Juno: magnitude 10.7. Virgo, not visible.)

07:29 UT - The just-past new Moon is at its farthest point from the Earth for the year. Distance from Earth: 406,465km (252,566 miles.) (Virgo, not visible.)

*September 15th*

04:16 UT – Dwarf planet Ceres is stationary prior to resuming prograde motion. (Magnitude 7.9. Sagittarius, evening sky.)

04:44 UT – The waxing crescent Moon is north of Mercury. (Mercury: 34% illuminated, magnitude 0.7, diameter 8.6". Virgo, evening sky.)

*September 16th*

00:35 UT – The waxing crescent Moon is north of Spica. (Virgo, not visible.)

Good opportunity to see Earthshine on the waxing crescent Moon after sunset. (Evening sky.)

*September 17th*

13:20 UT - Mercury is stationary prior to beginning retrograde motion. (28% illuminated, magnitude 0.9, diameter 9.0". Virgo, evening sky.)

The Sun leaves Leo and enters Virgo.

Mercury fades to magnitude 1.0. (28% illuminated, diameter 9.0". Virgo, evening sky.)

*September 18th*

Venus decreases its apparent diameter to 40.0". (24% illuminated, magnitude -4.5. Cancer, pre-dawn sky.)

Dwarf planet Ceres fades to magnitude 8.0. (Sagittarius, evening sky.)

*September 19th*

04:03 UT – The waxing crescent Moon is north of Saturn. (Saturn: magnitude 0.6, diameter 16.9", Libra, evening sky.)

21:23 UT – The waxing crescent Moon is north of Antares. (Ophiuchus, evening sky.)

*September 20th*

21:00 UT - Venus reaches maximum brightness. Magnitude: -4.55. (26% illuminated, magnitude -4.5, diameter 38.6", Cancer, pre-dawn sky.)

*September 21st*

08:59 UT - First Quarter Moon. This is the southernmost first quarter moon of the year. (Sagittarius, evening sky.)

*September 22nd*

10:48 UT – The just-past first quarter Moon is north of Pluto. (Pluto: magnitude 14.2. Sagittarius, evening sky.)

Asteroid 4 Vesta brightens to magnitude 6.0. (Cetus, evening sky.)

*September 23rd*

08:20 UT - Autumnal Equinox. Autumn begins in the northern hemisphere, spring begins in the southern hemisphere.

13:00 UT – The waxing gibbous Moon is north of dwarf planet Ceres. (Ceres: magnitude 8.0. Sagittarius, evening sky.)

Venus returns to Leo from Cancer. (29% illuminated, magnitude -4.5, diameter 36.9", pre-dawn sky.)

## September 24th

08:13 UT - Mars is 0.8° north of Regulus. (Mars: 97% illuminated, magnitude 1.8, diameter 3.9", Leo, pre-dawn sky.)

09:52 UT - Pluto is stationary prior to resuming prograde motion. (Magnitude 14.2. Sagittarius, evening sky.)

12:00 UT – Asteroid 4 Vesta is at perigee. Distance to Earth: 1.426 AU (Magnitude 6.0. Cetus, visible all night.)

## September 26th

10:06 UT – The waxing gibbous Moon is north of Neptune. (Neptune: magnitude 7.8, diameter 2.3". Aquarius, evening sky.)

Venus decreases its apparent diameter to 35.0". (31% illuminated, magnitude -4.5. Leo, pre-dawn sky.)

## September 27th

07:13 UT – Asteroid 4 Vesta has reached its maximum brightness for 2015. Magnitude: 6.0 (Cetus, visible all night.)

## September 28th

00:13 UT – Penumbral lunar eclipse begins.

01:12 UT – Partial lunar eclipse begins.

02:13 UT – Total lunar eclipse begins.

02:47 UT – Total lunar eclipse maximum. Visible from Africa, Antarctica, the Arctic, south and east Asia, the Atlantic, Europe, the Indian Ocean, most of north and south America and the Pacific.

02:50 UT - Full Moon. With a diameter of 33.472', this is the largest full moon of the year and the closest the Moon will be to the Earth for the year. (Pisces, visible all night.)

03:22 UT – Total lunar eclipse ends.

04:23 UT – Partial lunar eclipse ends.

05:22 UT – Penumbral lunar eclipse ends.

12:18 UT - Mercury is at perigee. Distance to Earth: 0.651 AU (2% illuminated, magnitude 4.2, diameter 10.3". Virgo, not visible.)

12:37 UT – Asteroid 3 Juno is in conjunction with the Sun. (Juno: magnitude 10.7. Virgo, not visible.)

21:47 UT – The just-past full Moon is at its closest to Earth for 2015. Distance from Earth: 356,876 km (221,752 miles.) (Virgo, visible all night.)

*September 30th*

14:36 UT - Mercury is at inferior conjunction with the Sun. Distance to Earth: 0.656 AU (0% illuminated, magnitude 5.1, diameter 10.2". Virgo, not visible.)

17:22 UT – Mercury is 5.3° south of asteroid 3 Juno. (Mercury: 0% illuminated, magnitude 5.1, diameter 10.2". Juno: magnitude 10.7. Virgo, not visible.)

# *October*

## *October 1st*

17:11 UT – The waning gibbous Moon is south of M45, the Pleiades open star cluster. (Taurus, pre-dawn sky.)

## *October 2nd*

13:45 UT – The waning gibbous Moon is north of Aldebaran. (Taurus, pre-dawn sky.)

## *October 3rd*

04:34 UT – Asteroid 4 Vesta is at opposition. (Magnitude 6.1. Cetus, visible all night.)

## *October 4th*

21:06 UT - Last Quarter Moon. This is the northernmost last quarter moon of the year. (Gemini, pre-dawn sky.)

## *October 7th*

Venus decreases its apparent diameter to 30.0". (39% illuminated, magnitude -4.5. Leo, pre-dawn sky.)

*October 8th*

20:41 UT – The waning crescent Moon is south of Venus. (40% illuminated, magnitude -4.5, diameter 29.8", Leo, pre-dawn sky.)

21:34 UT – The waning crescent Moon is south of Regulus in Leo. (Leo, pre-dawn sky.)

22:05 UT - Mercury is stationary prior to resuming prograde motion. (19% illuminated, magnitude 1.1, diameter 8.6", Virgo, pre-dawn sky.)

Mercury brightens to magnitude 1.0. (20% illuminated, diameter 8.6". Virgo, pre-dawn sky.)

The Draconid meteor shower peaks. Maximum zenith hourly rate: Variable. (Moon: waning crescent. Draco.)


*October 9th*

18:21 UT – The waning crescent Moon is south of Mars. (Mars: 97% illuminated, magnitude 1.8, diameter 4.0", Leo, pre-dawn sky.)

06:29 UT - Venus is 2.6° south of Regulus. (Venus: 41% illuminated, magnitude -4.5, diameter 29.4", Leo, pre-dawn sky.)

18:41 UT – Asteroid 3 Juno reaches apogee. Distance to Earth: 3.946 AU (Juno: magnitude 10.8. Virgo, not visible.)

22:58 UT – The waning crescent Moon is south of Jupiter. (Jupiter: magnitude -1.7, diameter 31.7", Leo, pre-dawn sky.)

Good opportunity to see Earthshine on the waning crescent Moon before sunrise. (Leo, pre-dawn sky.)

*October 10th*

Mercury brightens to magnitude 0.5. (25% illuminated, diameter 8.3". Virgo, not visible.)

*October 11th*

12:18 UT – Uranus is at perigee. Distance to Earth: 18.984 AU. (Magnitude 5.7, diameter 3.7". Pisces, visible all night.)

14:11 UT – The waning crescent Moon is south of Mercury. (Mercury: 32% illuminated, magnitude 0.3, diameter 7.9", Virgo, pre-dawn sky.)

18:48 UT - Uranus reaches its maximum brightness. Magnitude: 5.7. (Diameter: 3.7". Pisces, visible all night.)

*October 12th*

03:31 UT – The almost new Moon is south of asteroid 3 Juno (Juno: magnitude 10.8. Virgo, not visible.)

10:55 UT - Uranus is at opposition. (Magnitude 5.7, diameter 3.7". Pisces, visible all night.)

17:00 UT - Mercury brightens to magnitude 0.0. (37% illuminated, diameter 7.6", Virgo, pre-dawn sky.)

18:18 UT - Mercury is at perihelion. Distance to Sun: 0.308 AU (38% illuminated, magnitude 0.0, diameter 7.6", Virgo, pre-dawn sky.)

*October 13th*

00:05 UT - New Moon. (Virgo, not visible.)

05:18 UT – The just-past new Moon is north of Spica. (Virgo, not visible.)

*October 14th*

Saturn returns to Scorpius from Libra. (Magnitude 0.6, diameter 15.4", evening sky.)


*October 15th*

07:06 UT - Mercury is at half phase. (50% illuminated, magnitude -0.4, diameter 7.1", Virgo, pre-dawn sky.)

Mercury brightens to magnitude -0.5. (50% illuminated, diameter 7.1", Virgo, pre-dawn sky.)


*October 16th*

03:18 UT - Mercury is at Greatest Western Elongation. (54% illuminated, magnitude -0.5, diameter 6.9", Virgo, pre-dawn sky.)

11:55 UT – The waxing crescent Moon is north of Saturn. (Saturn: magnitude 0.6, diameter 15.4", Scorpius, evening sky.)

Best opportunity to see Earthshine on the waxing crescent Moon after sunset. (Scorpius, evening sky.)


*October 17th*

01:55 UT – The waxing crescent Moon is north of Antares. (Ophiuchus, evening sky.)

13:44 UT – Mars is 23' north of Jupiter. (Mars: 96% illuminated, magnitude 1.7, diameter 4.1". Jupiter: magnitude -1.8, diameter 32.1", Leo, pre-dawn sky.)

*October 18th*

19:26 UT - Mars occults the magnitude 4.6 star Chi Leo but will not be visible from the western hemisphere. (Mars: 96% illuminated, magnitude 1.7, diameter 4.1". Leo, pre-dawn sky.)

*October 19th*

20:58 UT – The almost first quarter Moon is north of Pluto. (Pluto: magnitude 14.2. Sagittarius, evening sky.)

*October 20th*

20:31 UT - First Quarter Moon. (Sagittarius, evening sky.)

*October 21st*

05:03 UT – The just-past first quarter Moon is north of dwarf planet Ceres. (Ceres: magnitude 8.4. Sagittarius, visible all night.)

Asteroid 2 Pallas fades to magnitude 10.0. (Ophiuchus, evening sky.)

The Orionid meteor shower peaks. Maximum zenith hourly rate: 25. (Moon: just-past full. Orion.)

*October 22nd*

Neptune fades to magnitude 7.9. (Diameter 2.3". Aquarius, evening sky.)

*October 23rd*

18:10 UT – The waxing gibbous Moon is north of Neptune. (Neptune: magnitude 7.9, diameter 2.3". Aquarius, evening sky.)

*October 24th*

Venus decreases its apparent diameter to 25.0". (48% illuminated, magnitude -4.4. Leo, pre-dawn sky.)

*October 25th*

06:30 UT - Venus is at half phase. (50% illuminated, magnitude -4.4, diameter 24.3", Leo, pre-dawn sky.)

*October 26th*

07:12 UT - Venus is at Greatest Western Elongation. (51% illuminated, magnitude -4.4, diameter 24.0", Leo, pre-dawn sky.)

08:16 UT – Venus is 1.1° south of Jupiter. (Venus: 51% illuminated, magnitude -4.4, diameter 24.0". Jupiter: magnitude -1.8, diameter 32.6", Leo, pre-dawn sky.)

09:32 UT – The almost full Moon is south of Uranus. (Uranus: magnitude 5.7, diameter 3.7". Pisces, visible all night.)

*October 27th*

12:05 UT - Full Moon. (Pisces, visible all night.)

Mercury brightens to magnitude -1.0 (89% illuminated, diameter 5.4". Virgo, not visible.)

*October 28th*

14:08 UT – Mercury is 4.2° north of Spica. (Mercury: 90% illuminated, magnitude -1.0, diameter 5.3". Virgo, not visible.)

*October 29th*

04:21 UT – The waning gibbous Moon is south of M45, the Pleiades open star cluster. (Taurus, evening sky.)

21:18 UT – The waning gibbous Moon is north of Aldebaran. (Taurus, evening sky.)

The Sun leaves Virgo and enters Libra.

Saturn brightens to magnitude 0.5. (Apparent diameter 15.2", Scorpius, evening sky.)

Dwarf planet Ceres fades to magnitude 8.5. (Sagittarius, evening sky.)

Asteroid 4 Vesta fades to magnitude 6.5. (Cetus, evening sky.)

# November

## November 1st

Mercury decreases its apparent diameter to 5.0". (94% illuminated, magnitude -1.0. Virgo, not visible.)

Mars enters Virgo from Leo. (95% illuminated, magnitude 1.7, diameter 4.3", pre-dawn sky.)

## November 2nd

Venus enters Virgo from Leo. (54% illuminated, magnitude -4.3, diameter 22.3", pre-dawn sky.)

## November 3rd

12:24 UT - Last Quarter Moon. (Cancer, pre-dawn sky.)

07:38 UT – Venus is 41' south of Mars. (Venus: 55% illuminated, magnitude -4.3, diameter 22.1". Mars: 95% illuminated, magnitude 1.7, diameter 4.3", Virgo, pre-dawn sky.)

## November 5th

01:31 UT – The waning crescent Moon is south of Regulus. (Leo, pre-dawn sky.)

## November 6th

17:16 UT – The waning crescent Moon is south of Jupiter. (Jupiter: magnitude -1.8, diameter 33.4", Leo, pre-dawn sky.)

Mercury leaves Virgo and enters Libra. (98% illuminated, magnitude -1.1, diameter 4.8", not visible.)

## November 7th

11:12 UT – The waning crescent Moon is south of Mars. (Mars: 95% illuminated, magnitude 1.7, diameter 4.3", Virgo, pre-dawn sky.)

15:57 UT – The waning crescent Moon is south of Venus. (Venus: 56.7% illuminated, magnitude -4.3, diameter 21.3", Virgo, pre-dawn sky.)

Dwarf planet Ceres leaves Sagittarius and returns to Microscopium. (Magnitude 8.5, evening sky.)

## November 8th

Good opportunity to see Earthshine on the waning crescent Moon before sunrise. (Virgo, pre-dawn sky.)

## November 9th

07:09 UT – The waning crescent Moon is south of asteroid 3 Juno (Juno: magnitude 10.9. Virgo, not visible.)

13:03 UT - The waning crescent Moon is north of Spica. (Virgo, pre-dawn sky.)

## November 11th

05:34 UT – The almost new Moon is north of Mercury. (Mercury: 99% illuminated, magnitude -1.1, diameter 4.7". Libra, not visible.)

17:47 UT - New Moon. (Libra, not visible.)

*November 13th*

00:47 UT – The waxing crescent Moon is north of Saturn. (Saturn: magnitude 0.5, diameter 15.1". Scorpius, evening sky.)

04:36 UT – The waxing crescent Moon is south of Antares. (Ophiuchus, not visible.)

Venus decreases its apparent diameter to 25.0". (59% illuminated, magnitude -4.3. Virgo, pre-dawn sky.)

*November 14th*

Good opportunity to see Earthshine on the waxing crescent Moon after sunset. (Sagittarius, evening sky.)

*November 16th*

02:18 UT – The waxing crescent Moon is north of Pluto. (Pluto: magnitude 14.2. Sagittarius, evening sky.)

22:26 UT – Asteroid 3 Juno is 6.4° north of Spica. (Juno: magnitude 10.9. Virgo, pre-dawn sky.)

*November 17th*

03:59 UT – Asteroid 4 Vesta is stationary prior to resuming prograde motion. (Magnitude 6.8. Cetus, evening sky.)

09:39 UT - Mercury has reached its maximum brightness. Magnitude: -1.3 (100% illuminated, diameter 4.6". Libra, not visible.)

14:54 UT - Mercury is at superior conjunction with the Sun. Distance to Earth: 1.446 AU (100% illuminated, magnitude -1.3, diameter 4.6". Libra, not visible.)

The Leonid meteor shower peaks. Maximum zenith hourly rate: 15. (Moon: waxing crescent. Leo.)

### November 18th

00:14 UT – The nearly first quarter Moon is north of dwarf planet Ceres. (Ceres: magnitude 8.6. Microscopium, evening sky.)

17:41 UT - Neptune is stationary prior to resuming prograde motion. (Magnitude 7.9, diameter 2.3". Aquarius, evening sky.)

Dwarf planet Ceres leaves Microscopium and returns to Capricornus. (Magnitude 8.6, evening sky.)

### November 19th

06:27 UT - First Quarter Moon. (Capricornus, evening sky.)

### November 20th

03:34 UT – The just-past first quarter Moon is north of Neptune. (Neptune: magnitude 7.9, diameter 2.2". Aquarius, evening sky.)

22:36 UT – Mars is at aphelion. Distance to Sun: 1.666 AU (94% illuminated, magnitude 1.6, diameter 4.6", Virgo, pre-dawn sky.)

Mercury leaves Libra and enters Scorpius. (100% illuminated, magnitude -1.1, diameter 4.6", not visible.)

### November 21st

00:42 UT - Mercury is at apogee. Distance from Earth – 1.450 AU (100% illuminated, magnitude -1.1, diameter 4.6". Scorpius, not visible.)

*November 22nd*

17:17 UT – The waxing gibbous Moon is south of Uranus. (Uranus: magnitude 5.7, diameter 3.6". Pisces, evening sky.)

Mercury fades to magnitude -1.0. (100% illuminated, magnitude -1.1, diameter 4.6". Scorpius, not visible.)

*November 23rd*

The Sun leaves Libra and enters Scorpius.

*November 24th*

Asteroid 2 Pallas leaves Ophiuchus and returns to Serpens. (Magnitude 10.1, evening sky.)

Jupiter increases its apparent diameter to 35.0". (Magnitude -1.9, Leo, pre-dawn sky.)

*November 25th*

01:57 UT - Mercury is 2.7° south of Saturn. (Mercury: 99% illuminated, magnitude -0.9, diameter 4.6". Saturn: magnitude 0.5, diameter 15.1". Scorpius, not visible.)

13:15 UT – The almost full Moon is south of M45, the Pleiades open star cluster. (Taurus, visible all night.)

17:54 UT – Mercury is at aphelion. Distance to Sun: 0.467 AU (99% illuminated, magnitude -0.9, diameter 4.6". Scorpius, not visible.)

22:44 UT - Full Moon (Taurus, visible all night.)

Mercury leaves Scorpius and enters Ophiuchus. (99% illuminated, magnitude -0.9, diameter 4.6", not visible.)

Asteroid 4 Vesta fades to magnitude 7.0. (Cetus, evening sky.)

*November 26th*

10:21 UT – The just-past full Moon is north of Aldebaran. (Taurus, visible all night.)

16:56 UT – Mercury is 3.4° north of Antares. (Mercury: 99% illuminated, magnitude -0.9, diameter 4.6". Scorpius, not visible.)

*November 28th*

10:53 UT – Venus is 4.5° north of Spica. (Venus: 66% illuminated, magnitude -4.2, diameter 17.8", Virgo, pre-dawn sky.)

Jupiter brightens to magnitude -2.0 (Diameter 35.3", Leo, pre-dawn sky.)

*November 29th*

08:54 UT - Venus is at perihelion. Distance to Sun: 0.718 AU. (66% illuminated, magnitude -4.2, diameter 17.6", Virgo, pre-dawn sky.)

22:00 UT - Saturn is at apogee. Distance to Earth: 10.992 AU. (Magnitude 0.4, diameter 15.1". Ophiuchus, not visible.)

Mars brightens to magnitude 1.5. (93% illuminated, diameter 4.7". Virgo, pre-dawn sky.)

Saturn leaves Scorpius and enters Ophiuchus. (Magnitude 0.4, diameter 15.1", not visible.)

*November 30<sup>th</sup>*

06:40 UT – Saturn reaches its minimum brightness for 2015. Magnitude: 0.4. (Diameter 15.1". Ophiuchus, not visible.)

07:36 UT - Saturn is in conjunction with the Sun. Distance to Earth: 10.992 AU. (Saturn: magnitude 0.4, diameter 15.1". Ophiuchus, not visible.)

The Sun leaves Scorpius and enters Ophiuchus.

# *December*

## *December 2nd*

13:30 UT – The nearly last quarter Moon is south of Regulus. (Leo, pre-dawn sky.)

## *December 3rd*

07:40 UT - Last Quarter Moon. With an apparent diameter of 29.754', this is the smallest last quarter moon of the year. (Sextans, pre-dawn sky.)

Uranus fades to magnitude 5.8. (Apparent diameter 3.6". Pisces, evening sky.)

## *December 4th*

03:10 UT – Venus is 2.6° south of asteroid 3 Juno. (Venus: 68% illuminated, magnitude -4.2, diameter 17.0". Juno: magnitude 10.9. Virgo, pre-dawn sky.)

03:42 UT – The just-past last quarter Moon is south of Jupiter. (Jupiter: magnitude -2.0, diameter 35.9", Leo, pre-dawn sky.)

## *December 6th*

00:46 UT – The waning crescent Moon is north of Mars. An occultation may be visible from some parts of the western hemisphere but details will vary depending upon your location. Check online for details. (Mars: 93% illuminated, magnitude 1.5, diameter 4.9", Virgo, pre-dawn sky.)

18:55 UT – The waning crescent Moon is north of Spica. (Virgo, pre-dawn sky.)

### December 7th

12:26 UT – The waning crescent Moon is south of asteroid 3 Juno. (Juno: magnitude 10.9. Virgo, pre-dawn sky.)

18:17 UT – The waning crescent Moon is north of Venus. (Venus: 69% illuminated, magnitude -4.2, diameter 16.6". Virgo, pre-dawn sky.)

Good opportunity to see Earthshine on the waning crescent Moon before sunrise. (Virgo, pre-dawn sky.)

Mercury leaves Ophiuchus and enters Sagittarius. (95% illuminated, magnitude -0.7, diameter 4.9", not visible.)

### December 8th

Mars increases its apparent diameter to 5.0". (93% illuminated, magnitude 1.5. Virgo, pre-dawn sky.)

### December 9th

Mercury increases its apparent diameter to 5.0". (94% illuminated, magnitude -0.7. Sagittarius, not visible.)

### December 10th

15:49 UT – The waning crescent Moon is north of Saturn. (Saturn: magnitude 0.5, diameter 15.1". Ophiuchus, not visible.)

*December 11th*

07:13 UT – The almost-new Moon is north of Antares. (Ophiuchus, not visible.)

10:29 UT - New Moon. (Ophiuchus, not visible.)

Venus enters Libra from Virgo. (71% illuminated, magnitude -4.1, diameter 16.2", pre-dawn sky.)

*December 12th*

15:24 UT – The just-past new Moon is north of Mercury. (Mercury: 92% illuminated, magnitude -0.6, diameter 5.1". Sagittarius, not visible.)

*December 13th*

07:33 UT – The waxing crescent Moon is north of Pluto. (Pluto: magnitude 14.2. Sagittarius, not visible.)

Asteroid 2 Pallas leaves Serpens and enters Aquila. (Magnitude 10.1, not visible.)

The Geminid meteor shower peaks. Maximum zenith hourly rate: 120. (Moon: waxing crescent. Gemini.)

*December 14th*

Good opportunity to see Earthshine on the waxing crescent Moon after sunset. (Capricornus, evening sky.)

*December 15th*

21:28 UT – The waxing crescent Moon is north of dwarf planet Ceres. (Ceres: magnitude 8.8. Capricornus, evening sky.)

*December 17th*

07:01 UT – The waxing crescent Moon is north of Neptune. (Neptune: magnitude 7.9, diameter 2.2". Aquarius, evening sky.)

*December 18th*

15:14 UT - First Quarter Moon. With an apparent diameter of 32.205', this is the largest first quarter moon of the year. (Pisces, evening sky.)

The Sun leaves Ophiuchus and enters Sagittarius.

*December 19th*

06:57 UT – The just-past first quarter Moon is north of asteroid 4 Vesta. (Vesta: magnitude 7.3. Cetus, evening sky.)

08:22 UT – Mercury is 3.9° south of Pluto. (Mercury: 84% illuminated, magnitude -0.6, diameter 5.5", Sagittarius, evening sky. Pluto: magnitude 14.2. Sagittarius, not visible.)

Asteroid 4 Vesta leaves Cetus and returns to Pisces. (Magnitude 7.3, evening sky.)

*December 20th*

02:31 UT – The waxing gibbous Moon is south of Uranus. (Uranus: magnitude 5.8, diameter 3.6". Pisces, evening sky.)

*December 21st*

02:17 UT – Mars is 3.8° north of Spica. (Mars: 92% illuminated, magnitude 1.4, diameter 5.2", Virgo, pre-dawn sky.)

10:48 UT - Mercury reaches its maximum brightness. Magnitude: -0.6. (81% illuminated, diameter 5.7", Sagittarius, evening sky.)

14:04 UT - Saturn is Antares. Separation: 6.2° (Saturn: magnitude 0.5, diameter 15.1", Ophiuchus, pre-dawn sky.)

*December 22nd*

04:48 UT - Southern Solstice. Winter begins in the northern hemisphere, summer begins in the southern hemisphere.

*December 23rd*

00:09 UT – The waxing gibbous Moon is south of M45, the Pleiades open star cluster. (Taurus, evening sky.)

18:12 UT – The waxing gibbous Moon is north of Aldebaran. (Taurus, evening sky.)

Venus decreases its apparent diameter to 15.0". (74% illuminated, magnitude -4.1. Libra, pre-dawn sky.)

The Ursid meteor shower peaks. Maximum zenith hourly rate: 10. (Moon: waxing gibbous. Ursa Minor.)

*December 25th*

11:11 UT - Full Moon. This is the northernmost full moon of the year. (Orion, visible all night.)

Uranus decreases its apparent diameter to 3.5". (Magnitude 5.8, Pisces, evening sky.)

*December 26th*

09:24 UT - Uranus is stationary prior to resuming prograde motion. (Magnitude 5.8, diameter 3.5". Pisces, evening sky.)

*December 27th*

Asteroid 4 Vesta fades to magnitude 7.5. (Cetus, evening sky.)

*December 28th*

Mercury fades to magnitude -0.5. (64% illuminated, diameter 6.5". Sagittarius, evening sky.)

*December 29th*

03:02 UT - Mercury is at Greatest Eastern Elongation. (60% illuminated, magnitude -0.5, diameter 6.7", Sagittarius, evening sky.)

*December 31st*

17:48 UT - Mercury is at half phase. (50% illuminated, magnitude -0.3, diameter 7.2", Sagittarius, evening sky.)

23:18 UT – The waning gibbous Moon is south of Jupiter. (Jupiter: magnitude -2.2, diameter 38.9", Leo, pre-dawn sky.)

Made in the USA
San Bernardino, CA
14 December 2014